Modelling the city

A fundamental concern of urban geography is how well residents are served by the city and how variations in well-being can effectively be monitored and evaluated. Traditional indicators of deprivation or service delivery, based on single areas and on published data, have proved of limited value.

Modelling the City examines the changing role of urban models in respect of both the need to re-address measures of urban well-being and the perceived need to bring model outputs more in tune with key planning problems. The authors argue that whilst there has been substantial progress with a wide range of theoretical problems in urban modelling, modellers have not paid enough attention to the usefulness of their model outputs in terms of indicators which offer new insights into the workings of the city or region. Too often in the past modellers have focused on the direct use of model predictions for planning without fully exploring the rich information base created from both data input into models and the outputs of model simulations. In addition, contemporary concern with performance measurement in many service organizations makes it appropriate to review the development of performance indicators in urban planning as a whole, and urban modelling in particular.

Modelling the City offers a new geography of performance indicators for the public and private sector based on the principles of spatial interaction. This new geography includes definition of both residence- and facility-based indicators. The spatial interrelationships between these two sets of indicators are likely to offer new insights into the age-old equity–efficiency problem.

C.S. Bertuglia is Professor in Urban and Regional Planning at Turin University, **G.P. Clarke** is Lecturer in Geography at the University of Leeds, and **A.G. Wilson** is Vice-Chancellor of the University of Leeds.

Modelling the city

Performance, policy and planning

Edited by
C.S. Bertuglia, G.P. Clarke and A.G. Wilson

London and New York

First published 1994
by Routledge
11 New Fetter Lane, London EC4P 4EE

Simultaneously published in the USA and Canada
by Routledge
29 West 35th Street, New York, NY 10001

Typeset in Times by
Solidus (Bristol) Ltd, Bristol

Printed and bound in Great Britain by
Biddles Ltd, Guildford and King's Lynn

British Library Cataloguing in Publication Data
A catalogue record for this book is available from the British Library

Library of Congress Cataloging in Publication Data
Modelling the city: performance, policy, and planning / edited by
C.S. Bertuglia, G.P. Clarke, and A.G. Wilson.
p. cm.
Includes bibliographical references and index.
1. City planning. 2. Urban policy. 3. Economic indicators.
4. Social indicators. 5. Performance. I. Bertuglia, Cristoforo Sergio.
II. Clarke, G.P. III. Wilson, A.G. (Alan Geoffrey).
HT166.M573 1994
307.76—dc20 93–37380

ISBN 0–415–09944–7

Contents

Figures

Tables

Contributors

Professor Cristoforo Bertuglia, Professor, Turin University, Architecture Faculty

Dr Mark Birkin, Research Director, GMAP, University of Leeds

Dr Graham Clarke, Lecturer, University of Leeds, School of Geography

Professor Martin Clarke, Managing Director, GMAP, University of Leeds

Dr Silvia Occelli, Senior Researcher, Institute of Socio-Economic Research for Piedmont

Dr Giovanni Rabino, Lecturer, University of Milan, Engineering Faculty

Dr Roberto Tadei, Lecturer, Turin University, Engineering Faculty

Dr Huw Williams, Senior Lecturer, University of Wales, Cardiff, Town Planning Department

Professor Alan Wilson, Vice-Chancellor, University of Leeds

Acknowledgements

The editors are extremely grateful for all the translation work undertaken by Angela Spence in Torino. She has meticulously translated all the Italian chapters into English and therefore saved the editors a great deal of time and hard work. Thanks Angela! Thanks also go to Penny Hartley and Rosemarie Temple for their significant efforts in typing and retyping parts of the manuscript and making sense of our many scrawls. Jane Senior acted as a coordinator in the final stages of the book and thanks go to her as well.

We wish to thank the Italian National Research Council (CNR) who provided financial support for the research described in this book.

1 Introduction

C.S. Bertuglia, G.P. Clarke and A.G. Wilson

In the relatively short history of urban and regional modelling there have been a number of key development phases. The first theoretical works in the 1950s and 1960s soon broadened into a second era of applied research ranging from subsystem work on new retail or housing sites to larger scale comprehensive models concerned with the impacts of major urban land-use changes. Spurred on by developments in computer technology this was the golden age for systems analysis and quantitative geography in general. Batty (1979, 1989) reviews this period in great depth.

The subsequent critique of these models and applications is now well known and documented in the literature. There were 'philosophical' critiques (Sayer 1976; Harvey 1989), criticisms of model design and use (Brewer 1973; Lee 1973) and what might now be termed 'critiques of fashion'; i.e. that model-based research was simply not the research direction many felt prepared to follow (nicely illustrated by Cosgrove 1989). Yet out of these critiques came new lines of research in urban modelling. Perhaps it was natural that such criticisms would force researchers to re-examine basic theoretical underpinnings. This new, distinctly theoretical stage brought major advances in optimization methods (Wilson *et al.* 1981), model design (Anas 1983) and urban dynamics, first in relation to subsystem models (Harris and Wilson 1978; Wilson 1981a; Clarke, M. and Wilson 1985a) and more recently in terms of comprehensive urban models (Bertuglia *et al.* 1987a, 1990a, 1993).

Whilst significant advances were made in the theory of urban models, a number of researchers began to rethink the applied (and planning) implications of such theoretical developments. This was partly fuelled by a number of complementary issues such as the increasing availability of a wide range of new data sources in both the public and private sectors (a problem which had dogged earlier applications) and the increased power and portability of personal computers bringing high quality computer graphics and presentation facilities to the desk-top environment of the potential user. However, it was recognized that better data and computers would not alone lead to more widespread applications. What was also required was a closer scrutiny of the *usefulness* of present model outputs. Too often in the past modellers had focused on the direct use of model predictions for planning without fully exploring the rich information base created from both data input into models and the outputs of model simulations. Increasing interest in these outputs has led to this volume on the definition and deployment of suitable 'performance indicators' from model-based research in a variety of planning contexts.

The central aim of this book is to develop and articulate a systematic and comprehensive framework for the conceptual identification and calculation of model-based performance indicators in a variety of public and private sector environments. Although the emphasis is more likely to be on equity or effectiveness in the public sector and efficiency in the private sector both applications will need to focus on indicators relating to households and individuals as well as indicators concerning facility location. The spatial interrelationships between these two sets of indicators are likely to offer new insights into the age-old equity–efficiency problem.

In Chapter 2 G.P. Clarke and Wilson review traditional indicator studies in urban planning and geography, particularly those arising from the work on territorial social indicators in the United States. It will be clear from this review that most indicators relating to system performance in some way have been calculated directly from published data and they are correspondingly limited. In the context of this book, models can usefully be seen as filling gaps in databases so that data and model outputs together constitute an expanded geographical information system which can then provide the basis for performance indicator calculations. In a sense, therefore, the rest of the argument can be seen as an attempt to provide a design for such an information system with modelling methods providing inputs and the outputs being performance indicators.

In Chapter 3 Bertuglia, Clarke and Wilson review the changing planning environment and the history of indicator use in model-based research, particularly that linked to the ideas of accessibility and consumer welfare. The message here is that it is possible to build on this early research to help articulate performance indicators which are more attuned to key planning problems of the day.

In Chapter 4 Bertuglia and Rabino reassess the importance of model-based research in contemporary planning by exploring how the 'scientific method' fits into the changing urban structure of modern cities. In particular, they are interested in the importance of performance indicators for evaluation. This places greater emphasis on defining indicators based on both efficiency and effectiveness. This theme is taken up by G.P. Clarke and Wilson in Chapter 5 who offer a 'new geography of performance indicators' based on a wide range of different urban subsystems. They review traditional indicators in each area of research and offer an alternative framework based on model outputs. The central argument is that a powerful new urban social geography is possible through the specification of residence- and facility-based indicators.

The emphasis in Chapter 6 is on more comprehensive urban models and their dynamics. Tadei and Williams review the progress in comprehensive models and offer new insights into the behavioural components of such models. They are particularly interested in how the outputs of such models can be related to consumer welfare or benefit measures (thus partly building on the indicators reviewed in Chapter 3).

In Chapter 7 Birkin focuses on indicators for the private sector using case material built up for a number of real-world applications in the retail sector. He expands on the basic indicators introduced in Chapter 5 to provide a more powerful set of residence and facility indicators for network planning and marketing.

The application of performance indicator research is examined in Chapters 8 and 9. First, Birkin *et al.* provide a range of examples of performance indicators for incomes, labour markets and retailing based on a wide range of applications undertaken at the University of Leeds. These are all examples derived from actual policy-related research. In Chapter 9 the emphasis is on a wider range of performance indicators for one particular Italian urban region. Occelli shows how such a package or suite of indicators can be effectively used to monitor the state of the city and region and to shed light on the variations in performance between its individual localities.

Concluding comments are offered in Chapter 10.

2 Performance indicators in urban planning: the historical context

G.P. Clarke and A.G. Wilson

2.1 INTRODUCTION

It has long been recognized that modern economies produce widespread disparities in individuals' or households' access to income across our cities and regions. This in turn leads to spatial variations in quality of housing, quality of environment and access to goods and services (both public and private). One of the most fundamental tasks of geographers and planners has been to recognize and measure the extent of these variations in the hope that budget holders and decision-makers could be influenced to channel resources in certain priority directions. Although such welfare measures have been criticized by some geographers and social scientists more interested in tackling causes than effects (see Section 2.5) there remains enormous interest in methods for quantifying and analysing such variations in 'quality of life' and using them in various contexts.

The aim of this chapter is to review the range of studies which have used economic and social indicators for measuring spatial variations in quality of life. This provides the platform for developing a more systematic and comprehensive framework for the conceptual identification and calculation of performance indicators for use in a variety of situations in urban planning and policy analysis. In Sections 2.2 and 2.3 we chart progress and problems associated with the so-called 'social indicator movement' from its basic origins in the 1960s to the present day. Examples are drawn from the United States and Britain in particular. The recent state concerns with performance indicators *per se* is reviewed in Section 2.4, and this is followed by a discussion of a number of important issues that a performance indicator focus raises (Section 2.5).

2.2 SOCIAL INDICATORS IN THE UNITED STATES

Great concern was being voiced in the United States during the 1960s about the lack of information on the social well-being of American individuals. Pure economic indices or indicators had been available for some time, often published annually in the form of Government reports. Whilst these were generally regarded as being useful and innovative, many commentators were concerned about the lack of similar indicators of *social* malaise. The US Department of Health, Education and Welfare (1969) remarked:

> It seems paradoxical that the economic indicators are generally registering continued progress – rising income, low unemployment – while the streets and the newspapers are full of evidence of growing discontent – burning and looting in the ghetto, strife on the campus, crime in the street, alienation and defiance among the young.
>
> (p. xi)

Research into social indicators was initiated primarily by NASA, who were concerned over the impacts, or secondary effects, of the space programme on American society. Indeed the first major book on social indicators (Bauer 1966) was financed by NASA, and this represents the initial drive for more accurate social reporting. At the same time, a former senator was putting forward his support for more information on social well-being (Mondale 1967a, 1967b). Mondale (1967b) describes the proposal for the 'Full Opportunity and Social Accounting Act' which called for a Council of Social Advisors with a joint congressional committee to carry out the social accounting by means of an annual social report. Similar conclusions and recommendations were reached by the National Academy of Sciences Social Science Research Council (1969) who called for improved social indicators that would measure the quality of life, particularly in its non-economic aspects:

> The Committee recommends that substantial support, both financial and intellectual, be given to efforts under way to develop a system of social indicators and that legislation to encourage and assist this development be enacted by Congress.
>
> (p. 6)

The Johnson Administration contributed significantly to the indicator movement with its publication 'Towards a Social Report' (US Department of Health, Education and Welfare 1969). Smith (1973) notes that this report represents the first real attempt to produce a social equivalent of the US annual economic reports. The basic aim of the report was to produce a set of social indicators, including health and social mobility, to 'satisfy our curiosity about how well we are doing' (p. xii).

Following Johnson, the Nixon Administration produced the National Goals Research Staff Report (1970) aimed at monitoring key goals or targets by selected indicators. After this, national Government reports were much less common. The major exceptions were the two substantial reports of the US Office of Management and Budget, 'Social Indicators 1973' and 'Social Indicators 1976'. However, a great deal of work had also been undertaken by other agencies and by many academics. For example, the Russell Sage Foundation commissioned the work of Sheldon and Moore (1968), which was concerned with sociostructural 'objective' indicators of social change, and of Campbell and Converse (1972), concerned with psychological or 'subjective' indicators of attitudes, expectations, aspirations and values (Carley 1981) (see Section 2.5 for more discussion on objective and subjective indicators).

It was the comparison of different cities across a country, however, that received most attention after 1970. As Murphy (1980) notes, this urban indicator research was simply an off-shoot of the social indicator movement, seeking to describe the social, economic and political conditions of metropolitan areas in particular. He, and also Flax (1978), provide a bibliography on a wealth of such urban indicator studies.

Lewis (1968) very much set the standard with a study which ranked the best and worst cities in the United States on a variety of indicators. The strategy was to be widely copied over the next few years. Louis (1975) and Liu (1975) for example both undertook similar strategies. Todd (1977) examined a hundred major cities using a set of eighty key indicators covering issues such as the economy, population, the environment, health, crime, education and recreation. The aim of Todd's study was typical, simply to provide a yardstick with which to compare conditions in one city with those in other cities.

The drive to focus more specifically on social indicators in a geographical context came from the work of David Smith. He recognized that advocates of social indicators had rarely been geographers and were typically not accustomed to thinking in spatial terms. Interestingly he reflects:

> It is the lack of maps more than any other thing that brings home to a geographer the very limited spatial perspective of the social indicators movement.
>
> (Smith 1973, p. 65)

Smith adopted the phrase 'territorial social indicators' to examine explicitly the role of space in social indicator studies. Whilst it was the geographer David Smith who introduced the terminology of the territorial social indicator, Knox (1975) makes an important observation:

> Territorial social indicators are not merely a product of the geographers' perspective on the general social indicator movement; they are a necessary and logical extension of any realistic system of social reporting. People live locally and experience the prosperity, stresses, expectations and satisfactions of their own locality. National social indicators are aggregates of these conditions and as such may mask important problems at the local level.
>
> (p. 11)

Both Smith and Knox highlight the opportunities for looking at problems at a number of different levels of resolution. In relation to territorial social indicators, Smith (1973) comments:

> It subsumes the concepts of 'local', 'regional', 'metropolitan' and 'urban' indicators, each of which can be regarded as a special category of territorial indicators.
>
> (p. 63)

Smith and his colleagues applied their ideas on territorial social indicators in case studies of Tampa and Gainsville in Florida (Smith and Gray 1972; Dickinson *et al.* 1972). The type of indicators used in the Tampa study are given in Table 2.1. Much of Smith's work was based on composite indicators, standardized for the purpose of amalgamation. In Dickinson *et al.* (1972) they used a linear scale transformation (on a scale of 0–100) and the transformed scores were summed and divided by the number of variables to produce a final 'quality of life' indicator.

Paralleling the work of Smith were many local reports on US cities aimed also at measures of social well-being and quality of life. Martin Flax (1978) provides a thorough review of studies appearing between 1970 and 1977. Indeed by 1977 most towns and cities in the United States had been

Table 2.1 Criteria of social well-being and variables used in the Tampa study

1 Economic status

(i) *Income*
1 Income per capita ($) of persons 14 and over, 1970
2 Families with income less than $300 (%), 1970
3 Families with income over $1,000 (%), 1970
4 Persons in families below poverty level (%), 1970

(ii) *Employment*
5 Unemployed persons (% total workforce), 1970
6 Persons aged 16–64 working less than 40 weeks (%), 1969
7 White-collar workers (%), 1970
8 Blue-collar workers (%), 1970

(iii) *Welfare*
9 Families on Aid to Families with Dependent Children program (%), October 1971
10 Persons aged 65 and over on Old Age Assistance (%), October 1971

2 Environment

(i) *Housing*
11 Average value of owner-occupied units ($), 1970
12 Owner-occupied units valued less than $10,000 (%), 1970
13 Average monthly rental of rented units ($), 1970
14 Rented units with monthly rentals less than $60 (%), 1970
15 Units with complete plumbing facilities (%), 1970
16 Deteriorating and dilapidated houses (%), 1971

(ii) *Streets and sewers*
17 Streets needing construction (% of total length), 1971
18 Streets needing scarification and resurfacing (% of total length), 1971
19 Sanitary sewer deficiencies (% of total area), 1971
20 Storm sewer deficiencies (% of total area), 1971

(iii) *Air pollution*
21 Maximum monthly dustfall (tons per square mile), 1969

22 Average suspended particulars 1969 (μg/m day), 1969
23 Maximum monthly sulphation 1969 (mg $SO_3/100\ cm^2$ day), 1969

(iv) *Open spaces*
24 Area lacking park and recreation facilities (%), 1971

3 Health

(i) *General mortality*
25 Infant deaths (per 1,000 live births), 1970
26 Death rate (per 10,000 persons 65 or over), 1970

(ii) *Chronic disease*
27 Cancer deaths (per 100,000 population), 1970
28 Stroke deaths (per 100,000 population), 1970
29 Heart disease deaths (per 100,000 population), 1970
30 New active tuberculosis cases (per 10,000 population), 1970

4 Education

(i) *Duration*
31 Persons aged 18–24 with 4 or more years high school or college (%), 1970
32 Persons over 25 with 8 years or less school (%), 1970
33 Persons over 25 with 4 years high school (%), 1970
34 Persons over 25 with 4 years college (%), 1970

5 Social order (or disorganization)

(i) *Personal pathologies*
35 Narcotic violations arrests (per 10,000 residents), 1971
36 Venereal disease cases (per 10,000 population), 1970

(ii) *Family breakdown*
37 Families with children, having husband and wife present (%), 1970
38 Persons separated or divorced (% ever married), 1970

(iii) *Overcrowding*
39 Dwellings with more than 1.0 person per room (%), 1970

(iv) *Public order and safety*
40 Criminal violation arrests (per 1,000 residents), 1971
41 Juvenile delinquency arrests (per 10,000 residents), 1971
42 Accidental deaths (per 100,000 population), 1971

6 Social belonging (participation and equality)

(i) *Democratic participation*
44 Registered voters (% population 18 and over), 1971
45 Eligible voters voting in mayoral election (%), 1971

(ii) *Equality*
46 Racial distribution index, 1970
47 Income distribution index, 1970

Source: Smith 1973, pp. 123–4

investigated in relation to community and local variations of well-being: typical are the studies of the Denver Urban Observatory (1973), the Urban Observatory of San Diego (1973), McNamara (1973) (for Albuquerque) and Monti (1975) (for Austin, Texas).

2.3 SOCIAL INDICATORS IN THE UK

It is widely agreed that the development of social indicators in the UK was very slow until publication of the first volume of *Social Trends* in 1970, when the idea of using informative indicators of social conditions received official sanction (Bracken 1981). The history of this publication is reviewed in Thompson (1978) and Carley (1981, pp. 114–17).

One of the first major UK studies was that of Shonfield and Shaw (1972). The collection of papers in this book provide an excellent introduction, not only to the general issues and terminology of social indicators (especially the papers of Cazes and Carlisle), but also in relation to specific subsystems, such as health and education.

The majority of UK studies which linked geographical issues explicitly to the social indicators were based on factorial analyses especially during the 1960s and 1970s. These built on the pioneering work of Shevky and Williams (1949) and Shevky and Bell (1955) in social area analysis research on US cities (especially Los Angeles) and Moser and Scott (1961) who classified all the towns and cities in the UK into statistical groups based on their social and economic differences or similarities. The typical format of these studies was to examine the similarities and differences between areas using the Census of Population. Giggs (1970) provides one of the first large-scale UK-census-based studies in his work on 'socially disorganised areas' in Barry, South Wales. He analysed a large number of census-based indicators, using principal components analysis, and followed this by the subsequent mapping of the various urban dimensions. Principal components analysis (which belongs to the wider field of factor analysis) uses input data to produce a correlation matrix which measures the relationship between all the possible pairs of variables. From the latent vectors of this correlation matrix components or factors are derived which group highly correlated variables from the input data. These are then open to scrutiny and interpretation. Davies (1970) explains this technique in more detail using data for Cardiff and Swansea. Amos (1970) provides a similar study of 'malaise' in Liverpool, again using a multivariate analysis to reduce a large number of inputs to more manageable dimensions of deprivation.

Since these early studies the census has continued to provide the backbone of intra-urban deprivation and quality of life studies. Craig and Driver (1972) more formally assessed the suitability of population census data for use as social indicators, under the identification and quantification of adverse social need in small areas. One of the most comprehensive studies of the mid-1970s was that of Holterman (1975). She looked at the concentration of housing stress using 'Small Area Statistics' of the 1971 population census. 'Concentration ratios' were designed representing the number of persons or households in those worst enumeration districts (EDs) with a particular form of deprivation as a percentage of persons or households in all urban EDs (Carley 1981).

Similar use has been made of the 1981 Census and will undoubtedly reappear using 1991 information. Davies (1984) uses seven basic indicators of 'deprivation' (standardized by a *Z*-score technique) to produce an eighth composite indicator or 'index of deprivation'. The top and bottom six wards of London are ranked accordingly (an analogy here with the earlier US inter-urban comparative studies). Similarly the work by Hirschfield and Rees (1984) and Hirschfield (1986) summarizes the progress made over a number of years on the development of a social information system concerned with the measurement of deprivation (for the Leeds area), based on the Censuses of Population for 1971 and 1981.

Apart from individual researchers, many institutions have undertaken intra-urban quality of life or deprivation studies using census-type indicators. For example, the Department of the Environment (1975) identified 'Housing Action Areas' (HAAs) and 'Priority Neighbourhoods' by interpreting the following household indicators (quoted in Bracken 1981, p. 284):

1 proportion of households without exclusive use of bath or shower, inside wc, hot water tap
2 proportion of households lacking one of the above amenities
3 proportion of households sharing one of the above amenities
4 proportion of households overcrowded
5 proportion of households in privately rented accommodation

HAAs involved about 3–4 per cent of the worst housing in Britain and gave the designating local authority the power to offer preferential house renovation grants and force improvements to be made, at the extreme by compulsory purchase (Carley 1981).

Similarly, the Planning Research Application Group (PRAG) at the UK's Centre for Environmental Studies (see Webber 1977, 1978, 1979) derived forty indicators from the Small Area Statistics of the 1971 Census and performed a comprehensive cluster analysis to aggregate residential types on the basis of their similarity. From these neighbourhood type classifications different areas were examined by reference to a variety of social indicators (see also Carley 1981, pp. 134–41).

Interest in community indicators also comes from the 'Community Indicators Programme' of the UK's Chartered Institute of Public Finance and Accountancy (CIPFA 1979). This followed a similar US scheme (the 'Social Economic Accounts System' developed by Fitzsimmons and Levy 1976) and was designed to publish social indicator data for UK communities and hence attempt to aid decision-making by local authorities:

> First, it is designed to facilitate the setting of priorities for resource allocation among various services in a local authority, both for policy and budget planning. Secondly, it assists in performance measurement – the study of efficiency or the ratio of inputs to outputs, and the study of effectiveness, which is the extent to which goals or objectives are met by service provision. Thirdly, the information produced by the CIPFA system provides ammunition for local government in their negotiations with central government for various grants like the rate support grant.
>
> (Kemp 1979)

These issues will be re-examined in later sections (see also the discussion in Anson 1991).

2.4 PERFORMANCE INDICATORS

All the studies mentioned above could, in some sense, be said to be implicitly concerned with the performance of a service. Indeed, in the quote by Kemp in the last section we see how the CIPFA study emphasized the importance of its role as measuring performance: the study of efficiency and effectiveness. However, in a UK setting, it has been the Government which has shown greatest interest in performance indicators *per se*. This interest stretches back over many years. Two early White Papers (Treasury 1961, 1967) both suggested new strategies for nationalized industries, calling for performance indicators to provide regular and systematic information on the success of each industry in controlling its costs, increasing efficiency and economizing in the use of manpower and capital resources. However, the main impetus to such monitoring has clearly come in the post-1979 Thatcher era, with key legislation in 1982 (the White Paper on efficiency and effectiveness in the Civil Service – Cmnd 2616) (see Lewis and Jones 1990 for an in-depth review). The drive for measuring performance in UK local authority service provision (overseen to a large extent by the Audit Commission) paralleled similar forces at work in the United States. Indeed, one of the most substantial studies of performance indicators in monitoring the effectiveness of municipal services comes from the Urban Institute in Washington, D.C. (Hatry *et al.* 1977), who suggest the following criteria for selection of performance indicators (summarized by Carley):

1. Appropriateness and validity: indicators must be quantifiable, in line with goals and objectives for that service, and be oriented towards the meeting of citizen needs and minimising detrimental effects;
2. Uniqueness, accuracy and reliability: indicators generally need not overlap, double counting should be avoided but some redundancy may be useful for testing the measures themselves;
3. Completeness and comprehensibility: any list of indicators should cover the desired objectives and be understandable;

4. Controllability: the conditions measured must be at least partially under government control;
5. Cost: staff and data collection costs must be reasonable;
6. Feedback time: information should become available within the time-frame necessary for decision-making.

(Carley 1981, p. 166)

Goldacre and Griffin (1983) note that these general concerns about performance indicators in nationalized industries and local government services were paralleled by a concern to assess the performance of the National Health Service (NHS) in the UK, and this organization has perhaps been subjected to greatest scrutiny. In January 1982 the Secretary of State for Social Services announced arrangements for departmental reviews of each Regional Health Authority, and the first list of indicators was published later in 1982 (see Goldacre and Griffin 1983; DHSS 1984). The first package of performance indicators came under the major headings of clinical activity, finance, manpower, support service and estate management. Table 2.2 lists a set of indicators concerned with clinical activity.

The review of health service indicators by Goldacre and Griffin (1983) also provides an excellent account of the importance of the performance indicator framework. They quote the Secretary of State for Social Services, from *Hansard*:

> Indicators will enable comparisons to be made between districts and so help ministers and regional chairmen at their annual review meeting to assess the performance of constituent health authorities in using manpower and other resources efficiently.
>
> (p. 3)

Table 2.2 Indicators of clinical activity in the National Health Service

1	Urgent, immediate or emergency inpatient admissions per 1,000 population served
2	All inpatient admissions per 1,000 population served
3	Average length of stay
4	'Throughput': average number of patients per head per year
5	'Turnover interval': average length of time a bed lies empty between admissions
6	Day cases as a percentage of deaths, discharges and day cases
7	New outpatients referred per 1,000 population served
8	Ratio of returning outpatients to new outpatients
9	Admission waiting lists per 1,000 population served
10	Estimated number of days taken to clear waiting lists at present level of activity
11	Percentage of obstetric admissions that result in births (including stillbirths)

Source: Pollitt 1985, p. 2

Similarly Yates (1982) makes an important point:

> One should not expect indicators about performance in the NHS to be absolutely precise or accurate, but that, although the use of indicators does not guarantee the identification of poor or good performance, they can be used to tell you something about the probabilities of where some elements of poor performance might be found.
>
> (quoted in Goldacre and Griffin 1983, p. 79)

However, not all commentators agree that a performance indicator framework for the NHS is the best strategy for further progress. The reviews of Klein (1982), Pollitt (1985) and Roberts (1990) all provide many reservations and lead us to look far more generally at the pros and cons of a performance indicator or urban indicator framework.

2.5 ISSUES AND PROBLEMS

Having discussed the ideas associated with social, urban and performance indicators, it is now necessary to address a number of important issues which emerge and in some senses must be borne in mind throughout this text. The first relates simply to how one determines the most important or appropriate indicators to study. This issue is not new: the Guillebaud Committee of the NHS in 1956 recognized the problem of finding the right statistics to measure efficiency (Guillebaud 1956). Similarly Knox (1978a) notes the difficulty of selecting indicators from the wide variety possible when using the Census of Population. He refers to the almost 'infinite number' which can be derived from the Small Area Statistics. This leads him to speculate:

> Notwithstanding the utility of carefully formulated source lists of variables, however, there remains the problem of selecting the most appropriate. With such a large number to choose from, the biggest danger is that of neglecting the more complex and apparently esoteric formulations in favour of conventional or more accessible (the two are often synonymous) variables which may not be as sensitive.
>
> (p. 77)

Gehrmann (1978) examined the way in which a widely different ranking of cities was possible simply through the choice of one particular social indicator system over another.

A related issue is that by concentrating on certain key indicators one might obscure equally pressing social problems (Henriot 1970). This might come about through sheer ignorance, lack of suitable information or data, or indeed, as some authors point out, through a deliberate policy to use social indicators as 'vindicators'. A quote from Brand (1975) is particularly apt here:

> The procedure is simple and well-known. When you are threatened by some unpleasant development, do a statistical appraisal of the situation.

> Unless you are extremely unlucky you will be able to get some figures which will justify you in doing what you were going to do anyway – often nothing. Even better, get someone in a university or research agency to do the study for you. You have a fair chance of knowing what he is going to say anyway so you can hand pick your chap and when he finally presents his report (usually two years late anyway, which is all to the good) it vindicates you.
>
> (p. 86)

The problem of 'vindication' is more real when indicators are ideologically loaded (Grichting 1984). The message here is that it is too ambitious to suppose that indicators can be defined whose values in *some precise sense* measure the scale of problems or the evaluation of policies. Rather, it is better to calculate a wide range of indicators which between them can be thought of as characterizing situations, usually with the addition of human interpretation. The basic principle, against this background, is this: the notion of 'indicators' should then be taken literally. That is, the set of performance indicators *offers an indication* of problems and effectiveness. When problems are identified in this way, it is usually valuable to mount a deeper investigation of their causes and possible solutions. This permits the introduction of more qualitative analyses where appropriate.

Having faced the problem of which indicators to use, we then face the task of judging whether a particular indicator is suggesting a 'good' or 'bad' performance or state. This leads on to issues of standards, norms or objectives. Goldacre and Griffin (1983, p. 11) explore this problem in relation to health care indicators and reflect that many authors point out that without objectives, standards, or at least a statement of expectations, there can be no appraisal of performance. Knox (1975) explains that 'absolute' indicators can be used where there is significant or substantial 'scientific' agreement over maximum or minimum levels necessary for certain aspects of well-being. His examples include minimum requirements of clean air and minimum levels of protein or intake (Knox 1975, p. 12). Often these 'scientific agreements' are contained within strategy plans. The Grey Book (DHSS 1972) for example, in relation to health care planning, emphasized that it is essential that performance is monitored in relation to health authority plans, i.e. the plan will provide the 'yardstick against which to measure performance'. Hence decisions about norms or standards are left to those at the top level of policy management.

However, more often than not, we can get round this problem by focusing on *relative* standards, which might themselves then be open to further analysis or discussion. These would normally relate to more general goals, or ideals concerning a fairer or more equitable society (terms which need more elaboration and will be returned to shortly). These sentiments are echoed by the progress report on the joint exercise between the DHSS and the Northern Region (DHSS 1982):

> It is recognised from the outset that no simple indicator or combination of

> indicators could lead to a firm conclusion on whether the use of existing resources was 'efficient' or 'inefficient', 'good' or 'bad'. Such a judgement could only be reached after further detailed study of local circumstances.
>
> (quoted in Goldacre and Griffin 1983, p. 77)

The issues of norms and standards and the possible solution of concentrating on relative values still raises questions concerning a more fundamental debate on whether the norms selected should reflect measures of efficiency or equity (or, more broadly, effectiveness – an argument we develop further below). That is, whilst it may be possible and useful to measure the performance of a system in terms of the efficiency of resource allocation and management, it may be more socially beneficial to explore what this means for the consumer in terms of his/her access and overall ability to consume against the price and availability of that consumption. Clearly both sides of the coin are important. In terms of efficiency Smith (1973) notes succinctly:

> We are all sensitive about the spending of our tax dollars, yet seldom do we ask for a cost–benefit accounting of this,
>
> (p. 57)

and in Smith (1977)

> The capacity to evaluate alternative locations accurately with respect to efficiency criteria is extremely useful. In the private sector it helps businessmen to make money or avoid a loss, which benefits those whose employment depends on the firm's viability and also customers whose satisfaction is determined to some extent by the price that they have to pay for the goods. In the public sector or in a system where the means of production are owned by the people, comparative cost-analyses can help to ensure that society gets the most out of its investment or that the production required is achieved at least expenses of public funds.
>
> (pp. 303–4)

However, concentrating purely on efficiency issues masks other important equity issues. The concern for equity in society is a theme echoed and widely discussed in all aspects of social science. In economics, 'welfare economics' is now a well founded subdiscipline especially concerning issues of distribution and resource allocation. Indeed Chisholm (1971) suggested welfare economics as a possible alternative to microeconomics as a point of departure for location theory. Bracken (1981) notes that proponents of the welfare view argue that one kind of welfare income measure is vital to a proper evaluation of the redistributional effects of urban policies (see Rawls 1973; Rowley and Peacock 1975; Walker 1980). Similarly Ben-Shahar *et al.* (1969) simply argue:

> The goal of the town planner is to make a plan that maximises the value of a social welfare function subject to a number of constraints.
>
> (p. 105)

There is now a large literature on equity and the distribution of facilities in fields such as sociology and social policy. Lineberry (1977) is perhaps one of the best introductions.

Geographers too have been equally concerned with issues of equity, welfare and social justice. Harvey (1973) provided much stimulus for a range of other texts during the 1970s as did the work on territorial social indicators mentioned earlier (i.e. Knox 1975; Coates *et al.* 1977; Smith 1977, 1979). Smith's philosophy built on Lasswell's (1958) definition of political science as the study of 'who gets what, when and how' and Samuelson's (1973) idea of economics as concerned with 'what, how and for whom' to add the important spatial dimension. Thus he defined human geography as the study of 'who gets what, where and how' (Smith 1974, 1977).

The solution to the efficiency versus equity issue will nearly always have to be some compromise. Both Harvey and Smith have stressed that it is not possible to ignore efficiency issues totally when in search of social justice:

> I want to diverge from the usual mode of normative analysis and look at the possibility of constructing a normative theory of spatial or territorial allocation based on principles of social justice. I do not propose this as an alternative framework to that of efficiency. In the long run it will be most beneficial if efficiency and distribution are explored jointly.... It is counter-productive in the long run to devise a socially just distribution if the size of the product to be distributed shrinks markedly through the inefficient use of scarce resources. In the long-long-run, therefore, social justice and efficiency are very much the same thing.
>
> (Harvey 1973, pp. 96–7)

> How quality of life is defined is, of course, subject to many further value judgements. . . It is sufficient here to say that it concerns efficiency in the use of resources and equity or fairness in distribution of the benefits and penalties of life.
>
> (Smith 1977, p. xi)

However, we should note that not all agree that efficiency and equity can be considered together. Culyer (1977) for example argues that the categories of justice and efficiency should be kept 'entirely separate' in case of 'logical inconsistency' and 'professional arrogance'. Pollitt (1990) argues, in terms of performance indicators, that most indicators in most systems are still proxies for economy and efficiency, and that effectiveness and equity are seldom captured. This is a major challenge in urban planning. New perspectives for dealing with this issue are outlined in Chapter 3.

Concern with equity issues has also led many researchers to criticize aggregate 'hard data' indicators (called 'objective' indicators, though, as we have seen, these too are often value-laden) and look instead towards indicators which reflect personal attitudes, beliefs and feelings: so-called 'subjective' indicators. The amount of interest in subjective indicators reflects a belief that the use of 'objective' measures alone is highly suspect, and researchers must go beyond 'objective' outputs and measure the 'reality' in which people live (see Schneider 1974; Angrist *et al.* 1976; Kennedy *et al.* 1978; Kuz 1978).

There are a number of variations within the subjective indicator literature. Andrews (1981) identifies three levels of such indicators: the 'general or global level', reflecting general happiness with life; 'particular life concerns', reflecting one's standard of living and evaluation of one's housing, family, neighbourhood etc.; and 'subconcerns', reflecting one's very local surroundings, even down to the scale of satisfaction with one's own kitchen! Hayden (1977) adds another important dimension: that subjective indicators should measure whether social arrangements, programmes and resource allocations are consistent with cultural values. Knox (1976) provides a list of typical 'domains' covered by social indicators (Table 2.3).

Whilst accepting that subjective indicators can provide useful signposts to social problems we must be aware of the fact that perceived levels of performance or service do not necessarily reflect actual levels. Carley (1981) gives three important reasons why this might be the case:

1. 'It is not feasible for clients to be dissatisfied with a service because of a variety of unmet needs which that service is not designed to fulfil.'
2. Lack of real knowledge might affect expressed feelings: 'For example, the majority of citizens might be quite satisfied with environmental health services which are in fact substandard and inefficient.'
3. Feelings may be purely influenced by attitudes to government and politics: 'For example, a conservative person may express that the level of police service is inadequate based on general feelings about "law and order" rather than his own experience of police or crime. A general rule about such misleading results is that they are possible whenever there are determinants of the subjective indicator other than the service under consideration' (Stipak 1979, quoted in Carley 1981, pp. 169–70).

However, despite these problems, subjective indicators may from time to time provide useful insights into real problems that should not be ignored. Again as Carley (1981) points out:

> If people perceive the streets as unsafe, or public transport as poor, then that is an issue even if the level of crime is low or the buses are frequent.
> (p. 170)

More recent work on subjective indicators has come from the University of Glasgow. Rogerson *et al.* (1989) ranked the towns and cities of the UK in terms of quality of life as perceived by their residents. Indicators include access to open space, price of housing, cost of living and access to services. Results showed that towns like Bradford in the UK, which would appear low down in league tables based on 'hard' indicators of social deprivation, appeared to be excellent places to live when judged by their residents.

Table 2.3 Illustrative 'subjective' indicators: weights assigned to ten life domains by 1,450 survey respondents in Britain using an eleven-point scale (0, completely unimportant; 10, overwhelmingly important)

Domain	*Score mean*
The housing conditions you live in (number of rooms, state of repair, provision of running water, sanitary facilities, garage, garden)	8.5
Your neighbourhood and its environment (friendliness, cleanliness, appearance)	7.9
Your state of health (freedom from illnesses and the availability of medical services if you do feel ill)	9.1
Opportunities and facilities for education (being able to go to well-equipped schools, colleges and so on)	7.6
Job satisfaction (how happy you are with the sort of work you do; how interesting it is)	8.1
Family life (being close to your family and relatives)	8.8
Your social status (what other people think of you; their respect for you in general; your standing in the local community)	6.1
Opportunities and facilities for leisure and recreation (things like parks, theatres, cinemas, sports centres and so on; and having the time to make use of these things)	6.6
A stable and secure society (a society without a lot of crime, vandalism and industrial strife, and one where you are taken care of if you are thrown out of work, or become ill, and when you retire)	8.5
Your financial situation (the sort of money you earn and the amount you are able to save if you want to)	7.9

Source: Knox 1976, p. 16

Finally, in relation to possible areas of concern with an indicator-based approach, it is important to be aware of the concern of some authors over the so-called 'fetishism of space': the belief that by concentrating on certain areas, or by identifying certain areas for aid or investment, one might miss other areas equally in need. Certain areas, then, may score relatively low on one aspect of welfare but not on other aspects (as O'Loughlin (1983) points out, areas of 'multiple deprivation' may be less extensive than previously thought), whilst conversely some areas of 'well-being' may contain pockets of deprivation which do not emerge from area-based studies. (This is often termed the 'ecological fallacy' of attributing average conditions to the entire population of the area under study.)

MacLaren (1981), for example, in a study of Dundee, found that deprived households were scattered throughout the city, whilst Glennester and Hatch

(1974) and Berthoud (1976) criticized area-based studies for neglecting large numbers of the deprived who do not happen to live in deprived neighbourhoods. Stegman (1979) too found that even a microlevel study at the 'block level' reveals significant variation within target areas, so that all blocks should receive an infusion of scarce resources. Clearly, it is important not to predefine areas of the city for special treatment. That is, indicators must be calculated for all regions so that such parcels are not missed or glossed over.

Some authors would go even further by arguing that deprived neighbourhoods are only surface manifestations of the 'withdrawal of capital and of unequal resource allocation' (O'Loughlin 1983). 'Radical' geographers, for example, hold that social structures themselves must be altered before meaningful change can be accomplished. Duncan (1974) argues that area-based studies are merely 'cosmetic', 'papering over the cracks' without touching the underlying causes of deprivation (see also Lee 1976; Hamnett 1979; Knox 1982, for a general discussion).

The response to this argument is fairly well known. Given that it is unlikely that such political or social reform is possible in the short run, solutions to current deprivation problems still have to be pursued and resources still have to be allocated and assessed. As Wilson (1984) stresses, 'to deny this, as some critics have, is a nonsense'. Similarly Knox (1985, p. 413) makes a simple point:

> Space may not be a major or independent social force but spatial outcomes are still of considerable economic, social and political significance.

Ideally different schools of thought each have a role to play.

> What is required is a dual programme of 'people policies' operating over the long term at the structural level to achieve a redistribution of wealth in society, and more immediate local level 'place policies' designed to improve the current position of the disadvantaged residents of urban Britain.
>
> (Pacione 1990, p. 201)

None of these problems can or should, of course, be glossed over. Indeed for those contemplating the use of an (area-based) indicator approach it is vital to be aware of the dangers and potential pitfalls. Yet as Henderson-Stewart (1990) argues, to claim that authors of indicator studies are not often aware of the limitations and pitfalls assumes a level of naivety which is unfair.

We have offered some responses to these criticisms because we remain optimistic that there is much useful ground still to cover. Carley (1981) reminds us that this is still a relatively young field and any further developments should be encouraged. In a sense, the rest of the book takes up that challenge.

3 Models and performance indicators in urban planning: the changing policy context

C.S. Bertuglia, G.P. Clarke and A.G. Wilson

3.1 INTRODUCTION

Having looked broadly at the use of indicators in urban and social geography in Chapter 2 the focus in this chapter is progress to date with indicators derived from model-based research. First, we briefly review the planning context for the first era of applied models (Section 3.2). This is followed by a recap on the importance and role of models in this period, focusing especially on model outputs (Section 3.3). Section 3.4 addresses problems associated with the first range of applications and advances in model-based theory since those days. This sets the scene for greater discussion on the contemporary role of models in planning (Chapter 4) and a new geography of performance indicators which are explicitly linked to a variety of urban and regional problems (Chapters 5, 6 and 7).

3.2 POST-WAR URBAN DEVELOPMENT AND PLANNING

Although much of the United States and Western Europe was already heavily industrialized by the onset of the Second World War, the period 1945–70 saw continued and rapid population growth across much of this region. The most densely urbanized countries saw new urban growth away from the traditional core areas of heavy industry. In the United States for example this second wave of urbanization occurred in the 'rim areas' of the northeast seaboard; Florida, Texas, Arizona and California (Miller 1973). In the UK, we have witnessed the growth of urban areas in the south and southeast at the expense of the core manufacturing regions of the north and Midlands (Lewis and Townsend 1987; Champion and Townsend 1990). Elsewhere in Europe the pattern is not dissimilar, with growth in newer areas and great problems of decline within an industrial corridor running from Turin and Genova in northwest Italy, through east and north France, the Saar and the Ruhr and south Belgium, to the Midlands and north of England and on to Glasgow and Belfast (Van den Berg *et al.* 1982; Cheshire and Hey 1988).

Planning response to this environment of population growth has clearly varied between countries although, after 1945, there was a growing trend towards intervention in city planning both in the United States and across most European countries (Burtenshaw *et al.* 1991). This intervention most commonly appeared first in the guise of some kind of town or city plan. The town plan normally mapped the future physical form of the town or city (often based on assumptions of linear growth), relying principally on the application of a series of location and land-use restraints. This kind of policy focused on prevention of development (or the attempt to prevent others from carrying out development). It often ignored interrelations between physical form and socio-economic mechanisms or the fact that any action which occurs in the city will inevitably generate further impacts. The wider implications of new developments were therefore often ignored and the plan operated in the conviction that it was possible to predetermine even the most detailed aspects of its future physical form.

The town plans of the 1950s were soon seen to be far too restrictive, particularly in relation to the emerging city-regions which were becoming more extensive with greater suburbanization. This led to a new era of 'structure planning' in many European countries (Hall 1982).

These new 'structure plans' covered larger areas than the old town plans and purely physical aspects took second place to the socio-economic aspects of development. They defined action to be taken, were instigative rather than regulatory, and dealt with the structure rather than details. The city, as it had traditionally been understood, was seen as only the central focus of a much vaster area which needed to be examined as a comprehensive whole. This complex whole began to be conceived of as a system. From the consideration of the city in isolation we had passed to the consideration of the *urban system* (for a more detailed treatment of this development see Bertuglia 1991).

The new era of structure plans, with their associated ideas on systems, coincided with the increasing availability of computers for analysis and prediction, providing an ideal 'test laboratory' for new methods of spatial planning. Batty (1979, 1989) gives an excellent account of the rise of urban modelling in such a planning environment. Following on from the first land-use and transportation models in the United States, urban modelling began to spread to Europe, with varying degrees of success.

As Batty (1989) explains, the object of many modellers was to 'both research the emerging "science" of urban modelling and to "apply" such science, often in the form of advice to strategic planning authorities'. In the UK, applications varied from 'subsystem' models such as shopping models (McLoughlin *et al.* 1966) to large-scale comprehensive models of urban development (i.e. Cripps and Foot 1968, 1970; Batty 1970; see also Foot 1981) based on the pioneering work on Pittsburgh by Lowry (1964). In Italy too, some large-scale models were developed, simulating the whole range of urban activities and population, disaggregated only where necessary. They were used for the analysis of the impact of single-sector and multisector policies. The projects included the Biella Plan (Bertuglia and Rabino 1975) for a large area in the north of Piedmont and the plan for Greater Turin

(Bertuglia and Rabino 1976) covering an area with a population of approximately 2 million people (1 million in the city of Turin and 1 million in its hinterland). These were followed by plans for thirteen other areas of Piedmont (Donna Bianco *et al.* 1981).

In the case of the Biella Plan in particular, the use of the model and the discussion of the results involved the participation of numerous public administrators mainly at municipal but also at provincial and regional level. The brief was to examine the implications of the subregional location and transport policies whose aim was to halt depopulation resulting from industrial decline. When it emerged from the study that the proposed policies would be likely to have the opposite effect, accelerating rather than slowing the outflow, a major political controversy was provoked. The modified policies suggested by the output of the model were nevertheless adopted and produced positive results. This experience provided a useful example of the capacity of a model to predict successfully counter-intuitive consequences of planning actions in a complex social and economic context.

In terms of model outputs, many of these applications focused directly on the use of model predictions in planning: population and job distributions, for example, and flows on transport networks. Urban modellers have not traditionally paid much attention to performance indicators as such. There are some notable exceptions and we pursue these below.

3.3 INDICATORS ASSOCIATED WITH EARLY MODEL-BASED APPROACHES

In this section we review the concepts associated with the early model applications that focused on the manipulation of model outputs. We begin with a brief exploration of the work of authors who have attempted to build social indicator models directly, thus relating this section more formally to Chapter 2. Then we examine the concept of 'accessibility' in model-based research, and we discuss the measurement of consumer surplus as a benefit indicator and other similar types of 'goal' indicators.

3.3.1 Social indicator models

Gross (1966) was one of the first to suggest that social and economic variables should be incorporated into social indicator models in order to explore the structure and performance of social systems. The 1970s saw many attempts to use an indicator-based framework in the construction of social models. Early exponents of such attempts were Sullivan (1971, 1974), Anderson (1972, 1973) and, in part, Land. Land (1971), for example, provided a critique of purely descriptive reporting of indicators, and argued instead for their incorporation within the 'conceptual system' of social processes. (For excellent reviews see Fox (1974), Land and Spilerman (1975), Warren *et al.*

(1980), Carley (1981) and Juster and Land (1981).)

From these texts it is apparent that there is a wide variety of social modelling approaches. A couple of examples illustrate this variety. One important researcher has been Ben-Chieh Liu working with the 'quality of life production model': for individual i, at time t, quality of life is measured as a function of physical (PH) and psychological (PS) inputs. That is,

$$\mathrm{QOL}_{it} = f(\mathrm{PH}_{it}, \mathrm{PS}_{it}) \tag{3.1}$$

Disaggregating further, the model form becomes

$$\mathrm{QOL}_{it} = f(\mathrm{EC}_{it}, \mathrm{PW}_{it}, \mathrm{EN}_{it}, \mathrm{HE}_{it}, \mathrm{SO}_{it}, \mathrm{PS}_{it}) \tag{3.2}$$

with the following definitions: SO, social inputs; EC, economic inputs; PW, political and welfare inputs; EN, environmental inputs; HE, health and education inputs. These are discussed in more detail in Ben-Chieh Liu (1978). Unfortunately the psychological inputs are held constant because these are not 'normally quantifiable at the present' (1978, p. 249). This is the inevitable operational breakdown when trying to incorporate subjective indicators into a modelling framework.

Land's framework has been based on 'sociometric' social models, introducing time series statistics aimed at dynamic quantitative modelling. The theme of this work is based on transition models, from the demographic work of Stone (1971, 1975) (inflow–outflow/input–output demographic accounting equations). The work is too detailed to be adequately described here but an excellent review appears in Land and McMillen (1981).

There is clearly much more ground to cover in social indicator model research and Carley (1981) explains some of the difficulties associated with such modelling. However, he does reflect that 'this type of approach is the only one which will allow for the continued development and refinement of social indicators in aid of social theory' (p. 85). Clearly this field will continue to provide much interest in the future.

The main theme of this section is models in geography and planning and how these might identify and monitor a variety of social and spatial problems. To tackle this we examine two of the most common types of model-based indicator study, beginning with accessibility and following this with consumer-surplus-based benefit measures and other such goal indicators.

3.3.2 Accessibility indicators

Geographers have long been interested in accessibility as a fundamental component of the discipline and to some extent as an indicator (see Chapter 2). Smith (1977) describes accessibility as basic to location planning whilst Pred (1977) and Knox (1978a) both note the importance of measuring 'quality of life' in relation to accessibility to key services and facilities. Similarly Koening (1980) concludes his paper:

> Accessibility appears more and more as a key concept in urban and transport planning. It expresses what is possibly the major function of cities: i.e. providing opportunities for easy interaction or exchange.
>
> (p. 169)

In view of its importance in geographical studies a number of authors have expressed surprise that few have addressed the 'crucial question of territorial variations' in terms of accessibility to opportunities and urban resources (Knox 1978b, p. 369). However, there have been some studies aimed at examining accessibility through the variables incorporated within or output from urban models. The most common practice has been to relate the concept to the basic gravity model or the broader family of spatial interaction models (Wilson 1971). Knox (1978b) explains why:

> Gravity models which are frequently used in geography and urban and regional planning to explain or predict spatial interactions of various kinds, are specifically designed to handle distance-decay effects, and can be easily modified to provide measures of accessibility which reflect more realistically the relative level of spatial opportunity inherent in any one part of an urban or regional system.
>
> (p. 373)

Hansen (1959) provided one of the simplest measures of accessibility in a gravity-like formula:

$$A_i = \sum_j E_j \exp(-\beta c_{ij}) \tag{3.3}$$

where A_i is accessibility in zone i to employment (E_j) facilities in zone j and c_{ij} is the cost of travelling from i to j (β is a parameter to be determined). This embodies the crucial idea that accessibility is related both to distance (or cost) and the scale of the opportunities at the distant locations. Schneider and Symons (1971) provide an early example of the direct use of the gravity model. They define an index of access opportunity (AO) as

$$\mathrm{AO}_i = \sum_j \frac{S_j}{t_{ij}^b} \tag{3.4}$$

where S is the size of some facility j, t is the time taken to travel from i to j and b is the distance decay parameter.

This standard measure can then be weighted according to other criteria, such as the rate of car ownership. Knox (1978b) incorporates parameters based on travel speeds by car and by transit (public transport) to transform the AO_i measure above to

$$\mathrm{TA}_i = C_i\left(\frac{A_j}{S_\mathrm{a}}\right) + (100 - C_i)\left(\frac{A_i}{S_\mathrm{t}}\right) \tag{3.5}$$

where TA_i is the new index of accessibility for zone *i*, C_i is the percentage of car-owning households in zone *i*, and S_a and S_t are the average times taken to travel a distance by car and transit respectively:

> The index then provides a reasonably sensitive yet robust indicator of the accessibility of different localities to a given spatial distribution of facilities.
>
> (Knox 1978b, p. 373)

There have been a number of modified versions of the above Hansen-type framework. Ingram (1971) for example looks at the 'Gaussian measure' to define

$$A_i = \sum_j S_j \exp\left(\frac{-d_{ij}^2}{v}\right) \tag{3.6}$$

where the variables are as above with *v* defined as a constant. But this, in effect, involves a specific choice of decay function and specific parameter value, both of which are difficult to justify.

Oberg (1976) develops the 'cumulative opportunities' measure:

$$A_i = O_i(D)\left(D - \sum_j \frac{d_{ij}}{O_i(D)}\right) \tag{3.7}$$

where $O_i(D)$ is the total number of opportunities available to household *i* within distance *D* from home. *D* is taken as a measure of maximum walking distance.

A retailing example is given by Guy (1977, 1983):

$$A_i = \frac{\Sigma_k d_{ij}^{\min}(k) E_k}{\Sigma_k E_k} \tag{3.8}$$

where $d_{ij}^{\min}$ is the minimum straight-line distance to a shop *j* in which good *k* is available and E_k is the mean expenditure per household on good *k*.

As Smith (1977) explains, once a satisfactory index of accessibility has been developed it can be used as a goal attainment indicator. Building on the Schneider and Symons (1971) example above, Smith develops an indicator of the 'efficiency of the present location of facilities, measured by accessibility to a given spatial pattern of potential customers'. By multiplying each place's AO index by its population (*N*), summing for all *i* and dividing by total population, Smith arrives at the following index:

$$A = \frac{\Sigma(AO_i \times N_j)}{\Sigma N_i} \tag{3.9}$$

It would be possible, of course, to replace AO_i by any other relevant measure within this formula.

In most of the above examples the gravity-type formula has been used without fully incorporating the kind of behaviour represented in a spatial interaction model. More generally, in terms of the family of spatial interaction models (Wilson 1971), it is possible to interpret the balancing factors, in the entropy-maximization derivation, as descriptive measures of accessibility (Wilson 1967, 1974). In the case of singly constrained models, the balancing factors are the inverse of the Hansen accessibility, and they can also be interpreted as playing a role representing 'competition' from other 'suppliers'. The doubly constrained model can be written as

$$T_{ij} = A_i B_j O_i D_j f(c_{ij}) \tag{3.10}$$

where

$$A_i = \frac{1}{\Sigma_j B_j D_j f(c_{ij})} \tag{3.11}$$

and

$$B_j = \frac{1}{\Sigma_i A_i O_i f(c_{ij})} \tag{3.12}$$

T_{ij} is the interaction between zone i and zone j, O_i is the origin total of interaction flows out of i, D_j is the total of interaction flows into zone j and c_{ij} is the cost of travel from i to j. The A_i and B_j terms may then be interpreted as Hansen-type accessibility indicators, each modified to account for competition (Wilson 1967).

Out of the work on these kinds of gravity and spatial interaction measures has come a variety of extensions and modified or alternative frameworks. Koening (1980), for example, describes the behavioural approach to accessibility measures, based on utility concepts. A suitable expression for a behavioural foundation of accessibility, applied here to destination choice, turns out to be

$$U^t_{ij} = V^t_j - c^t_{ij} + \varepsilon^t \tag{3.13}$$

where U^t_{ij} is the utility associated by an individual t, living in i, with a destination in j; V^t_j is the gross utility of achieving destination j for individual t; c_{ij} is the generalized travel cost or time from i to j for individual t; and ε^t is a random term. The problem then is to specify a suitable probability function for the random variable ε^t. Koening (1980) describes a number of different methods for achieving this, including the hivex method (Koening 1975) and the now well-known family of logit models (Domencich and McFadden 1975; Weibull 1976; Ben-Akiva and Lerman 1978). For a more general review of accessibility and random utility see Williams (1976) and Williams and Senior (1978).

All of these kinds of accessibility indicators became popular in the 1970s, especially in the field of transportation research where researchers were keen to examine the land-use impact of changes in the transport network. For more discussion and various alternative expressions of accessibility see, for example, Black and Conroy (1977), Davidson (1977), Burns (1979) and Morris *et al.* (1979).

There are at least two problems associated with any of the accessibility indices mentioned so far. The first is simply a restatement of an old problem: it is difficult to know how to weight them so that they can be combined. Experiment in a policy context is probably the only solution to this unless they can be converted into elements of consumer surplus. This last suggestion is achievable for the travel component, but not in relation to the benefits offered by the facility reached.

The second problem is tricky, but may be solvable. In measuring access to opportunities, no account is usually taken of whether those opportunities can be successfully taken up. Consider, for example, the Hansen measure of accessibility to employment in equation (3.3). A_i measures the access to total jobs, but should perhaps be modified to take into account the probability of actually getting one.

A modified index could be constructed as follows. Let X_i be the number employed in i and Y_i the number unemployed. Let x_i and y_i be weights. Then

$$A_i = x_i X_i \sum_j E_j \exp(-\beta c_{ij}) y_i Y_i \qquad (3.14)$$

If y_i is given a substantial weight, then the $\{E_j\}$ distribution may become more important than the $\{c_{ij}\}$ distribution in the spatial pattern of $\{A_i\}$.

3.3.3 Benefit indicators based on consumer surplus

Allied with the work on accessibility indicators have been other kinds of 'goal' indicators such as measures of spatial benefit, derived primarily in relation to welfare economics. The two sets of indicators, however, are often integral parts of the same study. Williams and Senior (1978, pp. 254–5) note that 'changes in accessibility indices are inextricably linked with changes in locational benefit accompanying a transport or land-use plan' (see also Gwilliam 1972).

One such spatial benefit measure derives from the consideration of consumer welfare. The argument follows that of Wilson (1974, 1976) and we can focus on the location of retail facilities as a specific example. Competitive equilibrium from the retailer's point of view is likely to be very different from the consumer's optimal location pattern, or welfare equilibrium. This was demonstrated by Hotelling (1929). The two equilibrium solutions are shown in Figure 3.1.

Figure 3.1(a) shows the welfare equilibrium with the two competing retailers at the first and third quartile points and hence consumers travelling

Figure 3.1 Equilibrium solutions

the minimum mean distances. Figure 3.1(b), however, shows the competitive retailers' equilibrium which has resulted from each retailer in turn moving nearer to the centre in order to obtain more than half the sales.

In real-world terms, the location of retail facilities will be a mixture of these two forces and a set of others: the unequal spread of spending power and the benefits of agglomeration in terms of increasing returns to scale, for example. To account for this, Wilson suggested that we should attempt to trade off the benefits against the costs of travelling to retail outlets by maximizing consumer welfare. To explore this trade-off we need to write the production-constrained shopping model (a more specific form of equations (3.10)–(3.12)) on which the analysis is based:

$$S_{ij} = A_i e_i P_i W_j^{\alpha} \exp(-\beta c_{ij}) \tag{3.15}$$

where

$$A_i = \sum_j W_j^{\alpha} \exp(-\beta c_{ij}) \tag{3.16}$$

and S_{ij} is the flow of revenue from zone i to zone j; e_i is the per capita expenditure on goods in zone i; P_i is the population in zone i; W_j is some measure of attractiveness (normally size) of retail facilities in zone j; c_{ij} is the cost of travel from i to j; α is a parameter that measures consumers' scale economies; and β is a parameter reflecting ease of travel. Wilson (1974) obtained a measure of benefit by writing

$$W_j^{\alpha} = \exp(\alpha \log W_j) \tag{3.17}$$

and rewriting (3.15) as

$$S_{ij} = A_i e_i P_i \exp\left[\beta\left(\frac{\alpha}{\beta}\log W_j - c_{ij}\right)\right] \tag{3.18}$$

It can then be seen that $(\alpha/\beta)\log W_j - c_{ij}$ represents the size benefits of shopping at j and c_{ij} the disutility of travel. Thus each consumer needs to maximize $(\alpha/\beta)\log W_j - c_{ij}$, and so the aim of planners might then be to maximize total consumer welfare Z where

$$Z = \sum_{ij} S_{ij}\left(\frac{\alpha}{\beta}\log W_j - c_{ij}\right) \tag{3.19}$$

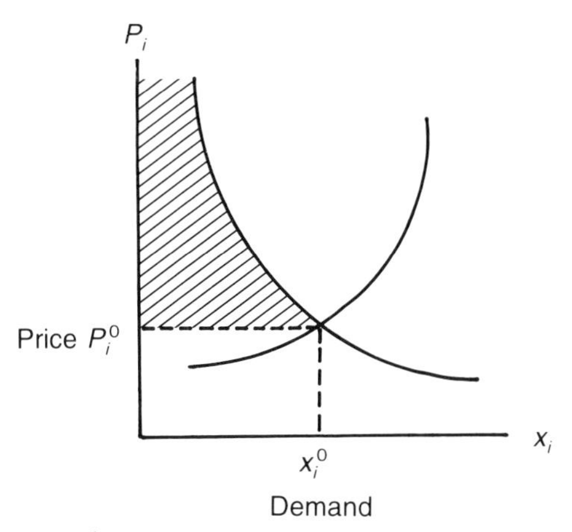

Figure 3.2 Consumer surplus

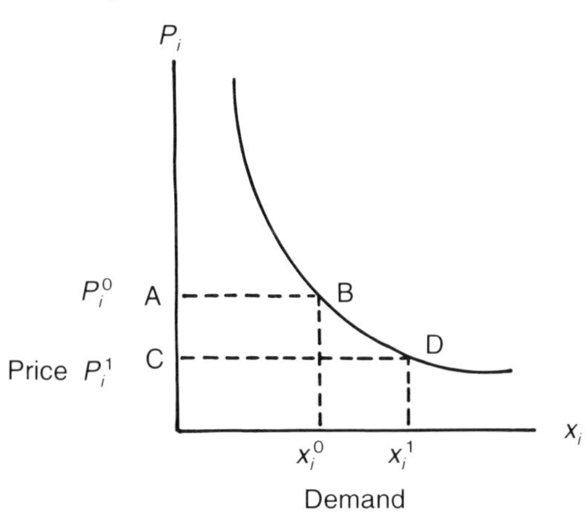

Figure 3.3 Change in consumer surplus

(subject to any number of additional constraints). Coelho and Wilson (1976) showed how this problem can be simplified and that the criterion in equation (3.19) is equivalent to the maximization of consumer surplus.

The idea of consumer surplus is a familiar one in economics. The shaded area in Figure 3.2 is known as consumer surplus since at any price greater than P_i^0, some people would still have been prepared to buy the good, and the shaded area represents the surplus of this group of consumers because the price is actually P_i^0 (Wilson 1974). Subsequent measures of benefit (B) thus arise from changes in consumer surplus (CS). If the price changes to P_i^1 then area ABDC in Figure 3.3 is the difference between the consumer surplus in the old situation and that in the new. From the area of the trapezium ABDC (after making the chord BD a straight line) the benefit (i.e. changing consumer surplus) which results from a price change is measured as

$$B = \Delta CS = \frac{1}{2}\sum_i (x_i^0 + x_i^1)(P_i^0 - P_i^1) \tag{3.20}$$

Wilson (1974, pp. 360–1) gives a simple example in relation to the transport sector, with a model that estimates T_{ij} (some flow total between zones i and j) as a function of c_{ij} (the cost of travel between i and j). If a transport facility reduces cost from c_{ij}^0 to c_{ij}^1 then the benefits of the new facilities are

$$B = \frac{1}{2}\sum_{ij}\left(T_{ij}^0 + T_{ij}^1\right)\left(c_{ij}^0 - c_{ij}^1\right) \tag{3.21}$$

(see also Tressider *et al.* 1968; SELNEC Transportation Study 1972; and more generally Neuburger 1971).

In terms of the shopping model introduced above (equations (3.15) and (3.16)), an equivalent maximization problem based on consumer surplus turns out to be (cf. Coelho and Wilson 1976)

$$\max Z$$

$$= -\frac{1}{\beta}\sum_{ij} S_{ij}\log S_{ij} + \alpha\left(\sum_{ij} S_{ij}\log W_j - H\right) + \beta\left(C - \sum_{ij} S_{ij}c_{ij}\right) \tag{3.22}$$

subject to

$$\sum_j S_{ij} = e_i P_i \tag{3.23}$$

and

$$\sum_j W_j = W \tag{3.24}$$

where the variables are as before and C is the total money available for travel; H is a constant.

This model could now be used to optimize the location and size of retail facilities with flows that are consistent with the spatial interaction shopping model (Coelho and Wilson 1976).

Williams (1976) and Williams and Senior (1978) argue that, for a 'wide class of spatial interaction models, computationally more efficient methods of generating the spatial benefits exist'. They look at the random utility maximization approach and focus attention on location surplus, the utility gained at a particular location minus the associated transport costs.

Accessibility and benefit indicators have been introduced here since they form the basis for many studies on land-use and transportation changes. It has also been recognized for some time that many other 'goal indicators' could be developed and we shall look at some of these briefly here.

3.3.4 Other 'goal' indicators

The earliest and perhaps most often used goal technique is cost–benefit analysis (see Mishan (1972) for a review). The aim is to weight the forecasted costs of undertaking a certain project against the expected benefits. Economic cost–benefit analysis tends to be primarily concerned with efficiency or variables based on income and finance, in contrast with social cost–benefit analysis which focuses attention on the well-being of members of society rather than decision-makers. Lichfield (1966) introduced the 'planning balance sheet analysis' (PBSA) which, crudely speaking, is a mixture of the two cost–benefit approaches. The aim of PBSA is to look at the effects of a wide variety of groups including members of the community and decision-makers.

A second well-known technique is the 'goal achievement matrix' after Hill (1968, 1973). Hill developed this procedure because of the limitations he felt were apparent in cost–benefit analysis and PBSA or, as Hill calls it, the development balance sheet. His main objection to these approaches was the lack of stated objectives. Hill (1968) remarks:

> Thus a major criticism of the 'development balance sheet' is that it does not appear to recognise that benefits and costs have only instrumental value. Benefits and costs have meaning only in relation to a well-defined objective. . . . It is meaningful to add or compare benefits and costs only if they refer to a common objective.
>
> (p. 21)

As regards his goal achievement matrix Hill comments:

> For the purposes of the goals-achievement matrix, goals should, as far as possible, be defined operationally, that is, they should be expressed as objectives. In this way the degree of achievement of the various objectives can be measured directly from the costs and benefits that have been identified.
>
> (p. 22)

Invariably, the information variables are 'weighted' in some preferred way and then summed so that the desired plan or policy is that which has the largest associated overall index or summation value.

In terms of urban modelling, Wilson (1974, p. 359) adapts Hill's goals achievement matrix as follows: identify goals or goal indicators by the index $g = 1, 2, 3, \ldots$ and associated weights α_g. Groups of the population are identified by $n = 1, 2, 3, \ldots$ and associated weights β^n. Costs and benefits can then be computed as C_g^n, B_g^n respectively for $n = 1, 2, 3, \ldots$ This gives Table 3.1. By adding the superscript m to indicate the mth plan, Wilson states the aggregate value indicator for plan m as

$$W^m = \sum_g \sum_n \alpha_g \beta^n (B_g^{nm} - C_g^{nm}) \tag{3.25}$$

Table 3.1 Hill's goals achievement matrix

Group n	*Group weight*	*Cost*	*Benefit*	*Cost*	*Benefit*
1	β^1	C_1^1	B_1^1	C_2^1	B_2^1
2	β^2	C_1^2	B_1^2	C_2^2	B_2^2
⋮	⋮	⋮	⋮	⋮	⋮

Source: Wilson 1974, p. 359
Note: Goal indicator $g = 1, 2, \ldots$; goal weight $\alpha_1, \alpha_2, \ldots$.

Thus the plan with the highest W^m can be chosen.

We shall return to the broader issue of evaluation in Chapter 4.

3.4 PROBLEMS AND ADVANCES IN URBAN MODELLING

The relative decline in the importance and use of urban modelling is a tale which is now well told (see Batty 1979, 1989). Broadly speaking there are three fundamental issues we wish to pursue here: the first relates to the sheer complexity of urban systems and the inability of large-scale models to be 'successful' (Brewer 1973; Lee 1973). In many instances this arose because advocates were either too optimistic or too ambitious in model design and perhaps too concerned with theory alone. In Italy, for example, it was recognized in retrospect that the 'novelty' of early applications lent perhaps disproportionate significance to the sheer innovation of the exercise. Although required to produce models for application to actual planning processes, there was a temptation to construct new models each time and to seek continually to refine the theoretical base. The consequence was a failure to exploit the potential of existing models fully and to provide them with the standard procedures which would have made them easier for the layman to use, less costly, and maybe more widely adopted. In retrospect we would say that given the circumstances, i.e. the need to provide models for practical application within a local authority context, the over-emphasis on theoretical and methodological aspects, more typical of university type research centres, should perhaps have been avoided. There were certain positive spin-offs, however. One was that it led to a better understanding of the complexity of urban systems and the nature of the processes of change.

Related to this issue was the changing 'culture of the time' (to use Bennett's phraseology (1989)). Batty (1989) outlines how the onset of the recession in the 1970s led to much of the Western world abandoning structure planning in favour of 'short-term tactical management' (p. 148). Hence, by 1980, structure planning was being dismantled across Europe for 'ad-hoc agreements and responses' (p. 163).

A second major setback to the development of applied urban models was the criticism from the wider social science community. The most notable attack was launched on modellers because of the undesirable association with positivism. This came from two directions. The humanists were concerned that the subjective experiences of individuals were being largely ignored, whilst 'radicals' or 'structuralists' argued that the assumptions of much (neo-classical) modelling were simply wrong, and that because modelling is positivist and functionalist it is likely to be concerned with surface observation symptoms rather than deeper causes.

The response to such criticisms has been to argue that there is a false association between positivism and modelling (Bennett 1985, 1989; Wilson 1989) and, perhaps more constructively, to work on a new phase of theoretical development. Undoubtedly there have been significant advances: Bennett (1989) refers to statistical work on spatial autocorrelation and spatial processes as well as models of dynamic systems of spatial structure and interaction. To this we might add new developments in extended comprehensive models (Birkin and Clarke 1985; Clarke and Wilson 1986; Bertuglia *et al.* 1990a, 1993) and new techniques such as micro-simulation (Clarke *et al.* 1980, 1981; Clarke and Holm 1987; Martin and Williams 1992). Despite these responses, it is clear that much of the spirit of the radical critique has to be accepted. We will seek to continue to respond in two ways: first, by continuing to deepen the theory which is at the basis of modelling (see the early work of Webber (1987) and Sheppard (1987), for example, as beginning to attempt to work Marxist analysis into economic modelling), and second by recognizing that there will always be a second investigative phase of planning work which goes beyond the model-based analysis. The conclusions can best be summarized in term of Habermas's three kinds of 'communicative interest': the 'technical', which is in this case connected to what can be achieved by a 'scientific' approach; the 'practical', concerned with interpretation and meaning in related discourse; and above all, the 'emancipatory'. The goal of all performance-indicator-based studies should be gains in emancipation of some kind. Habermas's second two 'interests' together imply that it will be necessary to recognize and rise above hidden ideologies in the frameworks which are developed. Modellers have to show that they can learn to be as capable of this as anyone else.

In practice, this will involve recognizing different categories of planning problems: those which are technical and relatively well defined; those which are not so well understood and which need interpretive studies; and those which involve power relationships in hidden ideologies. We will aim to show that the 'technical' and 'well defined', with appropriate extensions in modelling methods and insights, can provide appropriate foundations for planning studies in a greater variety of cases than is often thought.

The third key part of the attack on urban modelling relates to the *usefulness* of model outputs and this in a sense forms the heart of the rest of the book. Batty (1989) argues that models have generally produced too much data and that their outputs have been difficult to assess. In particular, there has been the problem of how to explain the complex mechanisms in the models to those not conversant with the techniques, and how to make it possible for the planning authorities to *compare* the results of the applications and the impacts of alternative policies or policy mixes which were often numerous and involved a great deal of complex data. Above all, how could they make these comparisons with reference to specific objectives or bundles of objectives?

This highlights the importance of finding a solution to the following problems:

1 how to convert the complex data output of the model into a manageable quantity of simplified but meaningful information (in other words, how to reduce a large quantitative output to a limited amount of qualitatively significant data);
2 how to present the fundamentally simulative applications of the model in a way which will facilitate the choice which the public decision-makers needed to make between the alternatives and provide the basis for the evaluation process.

These two problems have stimulated developments which have affected the whole conception of models themselves, methods of evaluation, the idea of performance indicators and the relationships between all three. We shall explore these in greater detail in Chapter 4.

The specification of new types of performance indicators needed the enabling technology to go with them. This was provided, or at least stimulated, in the 1980s by the arrival of the personal computer. First this has allowed powerful processing power to reside on the desks of a range of strategic planners (who, with some basic training, are no longer wholly reliant on the old fashioned computer departments of the past). It has in turn brought a new demand for software and a new environment of computer graphics which has transformed the way in which data and information (including model-based output) can be displayed and understood.

The demand for software extends not only to word-processing and database management packages. A new breed of 'planners' (both public and private sector) are demanding greater analytical power, and much of this involves understanding spatial distributions or variations in their businesses or planning activities. Openshaw (1986, 1989) and Beaumont (1987, 1989) are continually reminding us that there is no better time or opportunity for a new era of 'spatial analysis'. With appetites whetted by the early geodemographic systems (based on the manipulation of raw census data into area profiles) there is now massive interest in the role and use of geographic information systems (GIS): computer software packages which allow for the storage, manipulation and display of spatial data (see Rhind 1989). The success of GIS has in turn helped to rekindle applications of urban models. Purchasers (and users) of GIS are now realizing the very limited analytical power which is available in such proprietary packages. We argue elsewhere (Birkin *et al.* 1987, 1990, forthcoming) that what is needed is the power of analytical modelling to complement the power of data manipulation and display. This involves transforming traditional GIS into 'spatial decision support systems'

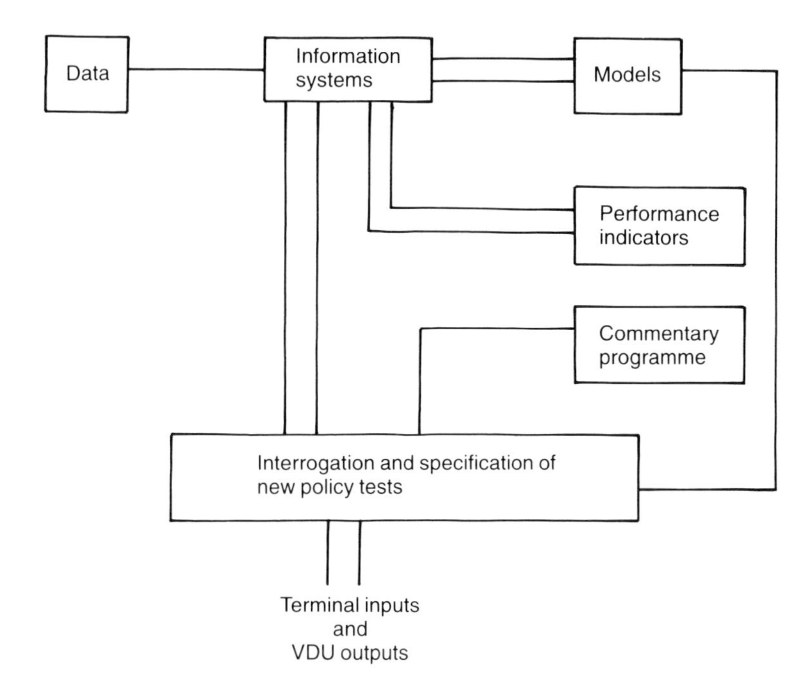

Figure 3.4 The role of information systems in urban planning
Source: Clarke M. and Wilson 1987

which provide descriptions (as a structural representation of urban data), impact analysis (as a method for measuring the effects of policy interventions) and evaluation (as a method for assessing the advantages and disadvantages of alternative choices). Since planning is usually concerned with medium- or long-term problems, there is also a role for indicators derived from dynamic urban models making it possible to build *adaptive information systems* which can react to urban change. This is now feasible as a result of the recent developments in decision support systems and artificial intelligence systems.

Figure 3.4 shows the key role of the modern information system in urban planning. It contains raw data and acts as a device for connecting models and performance indicators to database and mapping functionalities.

This new type of GIS should be connected to a programme which facilitates interactive interrogation and the specification of new policy tests. This could also be coupled with a commentary programme along the lines of that proposed by Clarke M. and Wilson (1987). Their ideas are taken further in Birkin *et al.* (forthcoming).

Although there is still scepticism that this route will bring modelling back into favour (Batty 1989) there is now much evidence that public and private sector planners are expressing renewed interest in what modelling can offer. Signs of such evidence will be offered in Chapters 8 and 9. Wilson (1989) is convinced that modelling has been rejected too soon by too many people. This may be so, but it is clear that modellers have not made themselves popular by failing to attune model outputs to key planning problems of the day. We shall return to this theme in Chapter 5.

4 Performance indicators and evaluation in contemporary urban modelling

C.S. Bertuglia and G.A. Rabino

4.1 INTRODUCTION

Performance indicators and simulation models (in our case those relating to urban systems) are instruments of both measurement and evaluation. As far as performance indicators are concerned, this double function is immediately evident. The idea of an indicator and the concept of measurement are obviously closely related, and clearly, the concept of performance also carries the direct implication of some form of evaluation. In the case of models these characteristics are reflected firstly in their function as a *quantitative representation* of reality, which presupposes a measurement of reality, and secondly as a *simplified representation* of reality. This too implies, consciously or otherwise, a choice (or evaluation) of variables and relations considered significant for the purpose of the model.

As a result of their common roots, there are a large number of relations, both logical and operative, between simulation models and performance indicators. These can be defined as

- complementary relations, which justify the joint use of the two tools (and often in fact make it useful to combine them, as for example in the reconstruction of missing data for model-based evaluation);
- relations of similarity, which impose conditions of logical and operative coherence.

The aim of this chapter is to attempt a systematic exploration of these relations. In the first part, we consider models as tools for measurement (for the quantitative representation of reality), recapping on the development of urban modelling in an attempt to outline the transformations which have occurred in our 'view of the world', i.e. the focus of interest and objectives pursued, and the implications that this has had for systems of urban performance indicators. We then move on to looking at simulation models as tools of evaluation, and therefore as performance indicators.

4.2 URBAN MODELLING IN A CHANGING WORLD

It is a widely held view that we are living through a period of profound and rapid economic, social and technological transformation. It is the opinion of many that these changes herald a new epoch. Cini for example (1990, p. 111 onwards) in an attempt to interpret the events occurring in the world today states:

> The real end of a century never coincides with its final years . . . the twenty-first century has already begun somewhere between the late nineteen-seventies and the early eighties. The twentieth century was the century of the working class, electricity, the dream of the future . . . suddenly the year two thousand is upon us, the working class no longer exists, the language of its ideology is no more, nor its vision of the future or its model of society . . . it is the passage from a linear way of seeing things to the awareness of complex chains of interactions and relations . . . it is not by chance that the twentieth century coincides with crisis in the governability of complex systems . . . it also coincides with the deposing of physics by biology as the exemplary science . . . and naturally it coincides with the computer boom.

Cini claims that what makes this change so profound is the *globality and interdependence* of the transformations affecting not only technology, economics and science, but society as a whole and our ethical systems. He also points to the *cyclical causal chain* (technology/society/science) and the cumulative nature of the resulting phenomena which explains why the change is so 'explosive' and radical.

It is possible that Cini's arguments (especially when presented in this highly summarized form) may not be totally convincing on a purely deductive level. They are in fact strongly coloured by a Marxian view, not shared by everyone, of the relationships between society and science and leave unresolved the problem of the logical priorities of the process of change (whether the leading factor is technology, science or society). It is nevertheless inescapable that in focusing attention on the relationship between science, society and technology Cini succeeds in identifying, at an analytical level, at least a conceptual structure which takes account *systematically* of all the main phenomena of the transformation taking place.

Exploring this structure (see also Rabino 1991), we cannot fail to be struck by the enormous increase in recent years in the rapidity with which scientific advance is being 'translated' into technological products and the growing 'scientific plus value' incorporated in these products. Technology on the other hand has been able in return to provide scientific research with increasingly sophisticated tools for investigation and measurement. We therefore find spin-offs in all sectors: innovations such as the introduction of new materials (from biotechnology), the availability of new instrumentation (from optonics) and a significant development in the potential of traditional products such as the car and the telephone achieved through the application of apparently minor modifications deriving from modern science. Similarly, all disciplines from the natural sciences to the humanities have been able to benefit from the availability of analytical and computational tools and instruments (developed for space exploration or high energy physics for instance) which provide ever increasing power and precision.

The distinction between science and technology is in fact becoming increasingly blurred. Technology frequently has to deal with problems of such a vast scale (e.g. worldwide communications networks) that the complexity and specialization involved make it hard to differentiate it from the activity of pure scientific research. Even more mundane technological issues when multiplied by the sheer scale or complexity of the problem or the number of people affected frequently become new scientific problems in themselves (as for example the problem of waste disposal or traffic in large cities). But science too is undergoing profound changes.

Technological progress has not only opened immense possibilities for study but has also radically altered the way in which research is carried out. It is sufficient to look at the potential offered for instance by modern methods of data processing to understand how revolutionary this has been. Fields which were previously impossible to penetrate such as the treatment of non-linear or discontinuous phenomena, disequilibria, or irreversible processes have now been opened. We have reached a point where we have to ask ourselves whether a revolution is occurring in the fundamental paradigms of scientific thought. It is this combined force of science and technology that confronts modern society (we refer above all, but not exclusively, to the so-called 'developed' countries). It is a society already profoundly transformed but still rapidly evolving. These changes have affected first its *demographic structure* as a result of the population explosion and increase in life expectancy made possible by modern medicine and increased living standards. There have also been significant changes in ethnic composition due to greatly increased geographical mobility. Second they have affected the *basis of social organization* deriving from the freedom from basic needs such as food and clothing made possible by the abundance of resources; the liberation from hard manual and repetitive labour towards freer and more creative occupations due to changes in manufacturing technologies and the organization of labour; the possibility of movement and interaction offered by modern means of transport and communications. Finally and above all, there is the *cultural dimension*. We only need to think of the sheer volume of 'information' available to everyone and reflect on the ever increasing level of education offered, and required, to participate actively in modern society to realize the extent of this transformation.

We are living in a far more profoundly and widely 'educated' (and hence more complex and differentiated) society which 'produces' science and technology. It is a society which needs this same science and technology for its existence and extols their value while at the same time it is conscious of the often negative implications. There is a growing ecological awareness and hence demands for a 'new science' which is more in tune with the 'new society'. The level of education and information so readily available has made modern society sensitive to the changes taking place and, in an attempt to understand where the change is leading, anxious to decipher the inevitably ambiguous and often contradictory signs. It is therefore a society which, reflecting on the past and aware of the present, looks to the future with great expectations and profound uncertainty.

The whole process of transformation which we have attempted to describe above is naturally reflected in the organization of human settlements and in particular the structure of urban and metropolitan systems which are the centres *par excellence* of modern society.

It is clear from the above discussion that it is the relationship between technological, scientific and social innovation and the city which is the crux of urban planning today, in relation both to the analytical aspects of urban science (Bertuglia 1991; Detragiache 1991; Rabino 1992) and planning policy formulation (Beguinot 1989). Urban modelling has shown itself capable of assimilating and reflecting these processes of change, in some cases more rapidly than other disciplines relating to the city. This is due, in part at least, to the use of the language of mathematics and a way of thinking which is more in tune with those sciences and techniques responsible for the transformation of society and the city.

In the last chapter we briefly discussed the evolution of urban modelling. The transition from first-generation (Lowry-type) models to second-generation models such as those formulated by Allen *et al.* (1978), Lombardo and Rabino (1983) and Bertuglia *et al.* (1990b) could be seen, in Cini's terms, as a move from modelling for the twentieth-century city to modelling for the twenty-first-century city. We do not intend to imply that the Lowry model (1964) represents an obsolete view of the city. In fact it provides a logical analysis of the city which is still valid today, but the kind of interpretation it provides (with the emphasis on the 'causal mechanisms' of the model) and its operative implementation (the interpretation of the relations between residence and employment are seen almost exclusively as journey to work trips, i.e. physical movements) recalls the image of the 'workers' city. Linked to this view is the conception of the city as a machine which can be steered towards desired objectives with exogenously defined actions and the idea of the model as an instrument whose main purpose is to provide a more rigorous way of sounding out the cause–effect relations.

Similarly we do not imply that second-generation models relate strictly to the city of the year 2000. They do capture, however, consciously or otherwise, certain aspects of the transformation towards the new kind of city. The model of Allen *et al.* (1978) for example focuses on the problem of self-organization and irreversibility of urban dynamics; that of Bertuglia *et al.* (1990b) concentrates on the synergies operating in the micro–macro interrelations, and the model of Mela *et al.* (1987) highlights the number of spatial organization principles present in the modern urban and metropolitan structure. It is perhaps this last model which deals most explicitly with the relationships between the model and transformations occurring in modern urban settlements today.

4.3 TOWARDS A NEW VIEW OF THE CITY AND THE PLANNING PROCESS

A natural extension of the changes described above was the conception of the city as an open system with many parts and relations, in non-equilibrium, with numerous decision-makers generating non-linear dynamics and therefore a vast range of possible futures. The systems view emphasized the possibility of discontinuous, and often irreversible, change in both time and space, and the capacity of the city for self-organization. For a more detailed treatment of the subject see Bertuglia (1991) and Bertuglia and La Bella (1991).

This development was eventually to help the local authorities adopt a more appropriate approach to the problem of planning modern urban systems and to extract themselves from the blind alley, and consequent impotence, into which the traditional planning system had led them. It became apparent that the qualitative changes that were occurring gave a new crucial role to the question of *evaluation*. There began to be less emphasis given to the search for *the* future configuration. The uselessness of attempting to define and then impose one given form on the city became evident. Discussion started to focus around the concept of possible future 'trajectories' and the consequent measures necessary to 'nudge' the system along a more desirable path. It became evident that it was essential to have at all times a complete panorama of the possibilities (positive or negative) open to the system, in order to be able to decide *what to do here and now*, bearing in mind the need constantly to review the set of future decisions and always concentrating on the needs of the present.

An approach such as the one described above involves not only a radical redefinition of the whole planning system and its tools but also a more central role for the process of evaluation and the need for more sophisticated and efficiently organized evaluation methods. From this it follows that we also require a systematic method for monitoring the evolution of the system and its future prospects, and hence the relevance of performance indicators.

First, however, let us look in more detail at our ideas on evaluation.

4.4 DECISION-MAKING AND EVALUATION

To be able to adopt a correct approach to evaluation in urban planning it is necessary to have a clear idea of the characteristics which have come to typify decision-making in the urban context. We should like to focus here on some of the more important features of this process.

First, the *conflictual nature* of the problems is becoming increasingly evident. Conflicts may arise for example between interest groups when attributing priorities to the objectives (e.g. economic growth versus environmental conservation), between different areas or authorities (the notoriously difficult job of finding sites for 'undesirable' activities such as waste dumping) or between different time horizons (where exploitation of a limited resource for the present generation has negative consequences for future generations). There may well be intra-personal conflicts for the decision-maker himself or herself (who could find contradictions for example between his or her own lifestyle and the environmental issues for which she/he is pressing). The result is that not only will different decision-makers probably be pursuing different objectives, but even an individual decision-maker may find that he/she has to weigh up a number of conflicting objectives.

Second, it has become increasingly common for decisions to be taken not by one individual, but by a consensus. To be more precise, decisions are now more and more frequently the outcome of an *institutionalized process* with predetermined procedures, in the course of which a whole series of decision-makers are required to intervene or be consulted. This adds the further complication that such a process requires time and the decision-maker may modify or even change his or her objective, or objectives, during the decision-making process.

Lastly, given the growing complexity of urban systems, it is becoming increasingly clear that a given problem will have not one obvious solution but a whole *range of possible alternatives*.

To summarize we can say that at the scale of the urban system any decision-making activity will have the following characteristics:

- multi-objectives;
- a complicated and therefore time-consuming decision-making process during which decision-makers may modify their views;
- a wide range of possible solutions.

The observations made above explain the need to reflect carefully on the whole concept of evaluation. Although there is often a tendency to emphasize one to the exclusion of the other, evaluation involves two inseparable elements. These are measurement and comparison. Evaluation can in fact be defined as the identification, and therefore the measurement, of the effects of an action (in this case the impact on a given urban system) in relation to a given set of objectives and constraints. It may also involve the comparison between the effects of one action and those of alternative actions, always in relation to stated objectives. As explained above, a comparison cannot be made if elements of measurements are not available.

Measurement and comparison are complex operations which pose both theoretical and practical problems. They have been the subject of an extensive body of studies and are still highly controversial issues (see again the discussion in Chapter 2). The most recent and advanced tools to have come out of this research are performance indicators and multi-criteria methods, which provide means of tackling the problems of measurement and comparison respectively. The former basically provide a way of transforming the raw data into meaningful information which can be used for measurement and comparison. The latter on the other hand provide a way of obtaining a single ranking which will permit the choice of the preferable alternative from evaluation of the various impacts (see Voogd 1983 for example).

The fact that performance indicators and multi-criteria methods are solutions to different but interconnected problems is the reason for their frequent combined use. To put this in practical terms, it means that performance indicators can be used as inputs into multi-criteria analysis. Problems may arise, however, when these indicators are calculated from the output of mathematical models which are formulated as optimization problems, even when they are basically simulation models (Wilson 1974; Boyce and Southworth 1979). For a more detailed examination of the problems arising from the combined use of performance indicators and multi-criteria methods we refer the reader to Bertuglia *et al.* (1988, 1989b, 1991b). We now wish to look at certain conceptual implications of the evaluation process in greater detail.

4.5 EVALUATION IN URBAN MODELLING AND PLANNING

According to Camagni (1988) and Mela and Preto (1990) a useful definition of the field of planning is given by the combination of two variables (Figure 4.1). The first concerns the subjective dimension, i.e. the decision-makers themselves and the kind of decision-making process, and the second the objective dimension of the plan, i.e. the nature of the system being planned.

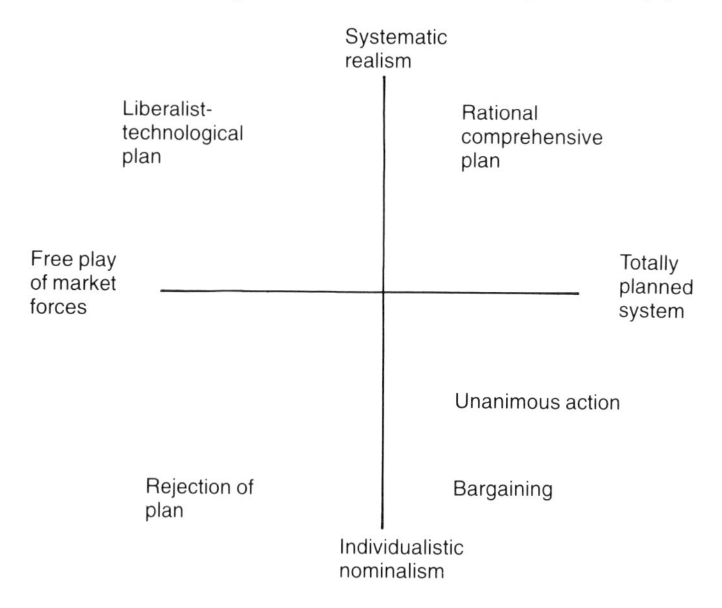

Figure 4.1 Possible conceptions of planning (and evaluation)

The first variable concerns the degree to which the decision-making system is 'coordinated'. At one extreme is the situation in which independent decision-makers have total autonomy in their own fields and hence the urban system is left at the mercy of the various mechanisms which affect it, directly and indirectly, as market forces have free play. At the opposite extreme we have total coordination, where planning decisions, taken by agreement or through a hierarchical structure, guide all aspects of the urban system (complete planning control). Naturally, the degree of coordination between decision-makers can vary considerably (some forms being more restrictive than others), leaving a wide range of intermediate possibilities between the two extremes.

The second variable concerns the degree of 'objectivity' of the model in reflecting or 'interpreting' what exists in nature. At one end of the scale is the view that the physical, economic and social entities making up the urban system have an objective reality, whose essence and relations can be explained by 'strong' theories (we call this view of reality 'systematic realism'). At the other end of the scale, the 'weak' concept of thought considers these entities as nominal categories belonging to an elusive complex reality, impossible to describe with precision and identifiable, if at all, only through metaphor or analogy (we have termed this 'individualistic nominalism'). Here too a whole range of intermediate positions exists, depending on the degree of emphasis given to the observer's role and his or her way of conceiving reality.

As we can see in Figure 4.1, all the various views on planning (and hence evaluation) can be 'located' by considering their relationship to these two variables: from the complete negation of coordinated planning activity to the totally comprehensive plan with a 'strong' decision-making structure; from the plan based on negotiation and agreement between decision-makers to the 'adaptive' plan where urban systems are left to the powerful but uncoordinated influence of market forces.

It is interesting to look at where performance indicators and models, as evaluation tools, fall in relation to this diagram. Because of their interpretive function, simulation models will obviously lie close to 'systematic realism'. This is particularly true for comprehensive and highly disaggregate models where the claim to globality means that the space left for the 'unexplored complexity of reality' is necessarily reduced to a minimum, and usually discounted as 'external' to the model. A high degree of disaggregation implies, often in conjunction with the need for quantification (to make the model operative), a detailed description of concepts and therefore greater use of laws or empirical hypotheses to explain the relationships between variables.

It does not follow, however, that more partial or aggregate models (for single parts of the urban system or relations between certain variables) are necessarily non-systematic. Where a model involves a specific aspect of the city for which the analytical approach adopted is likely to be less controversial (one of the more 'physical' components for example), it may well reflect a highly systematic view. Similarly the fact that a model is technically less refined, for reasons of limited availability of information or because a more elementary mathematical technique is employed, does not automatically mean that it is based on a non-systematic view of reality. The degree of sophistication or validity of a model or its underlying theories may influence its effectiveness as an evaluation tool, but the model–reality relationship is another question altogether. Single performance indicators on the other hand (or suites of indicators not systematically interrelated) seem in general to reflect a less rational/systematic view.

In assessing the underlying conceptual standpoint it is important not to base one's assumptions simply on the joint use of a formalized model. The formulation of an indicator for an evaluation problem already presupposes the construction of a logical/conceptual system of reality, since it is this which permits the definition of the problem and makes it possible to correlate the indicator. The positioning of the indicator in our diagram therefore is determined purely by the conceptual view it reflects and has no connection with the adoption or otherwise of a mathematical model.

Another highly significant aspect of the evaluation process is the way in which the evaluation criteria are translated into operative terms. These criteria are often in fact expressed in very general abstract terms (e.g. well-being, health, feasibility etc.) without specifying their implications on the structural details of the system. The construction of the performance indicator on the other hand requires these implications to be made more precise. This gives rise to a number of problems including

- the need to use a number of indicators to relate to one particular criterion;
- the impression of 'over-simplification' or inadequacy of the indicators in expressing the real intent of the evaluation criteria;
- the emergence of situations in which, in extreme cases, the same indicator can be interpreted in two completely opposite ways.

It is also useful to make some observations about the joint use of simulation models and performance indicators. We are referring here specifically to indicators which are 'derived' from the model and not to situations in which indicators are used to evaluate aspects external to the model. (In this case the comments already made above apply. In fact when the model is not fully comprehensive and the indicators are not included, it automatically implies that it has been impossible to relate the treatment to a systematic approach.) It is useful in this context to recall the need for logical coherence in the evaluation, as this is not always implicitly guaranteed in the derivation of the indicators. Although we are dealing with a simulation we should not forget that there is always an underlying theory, and the indicators should not be in contrast with the principles or optimization assumptions implicit in that theory. For example, if effectiveness or equity indicators are associated with spatial interaction structures, and these are described by entropic or gravitational type models (or probabilistic/behavioural models), the indicators cannot be evaluated without taking into account the implicit 'random' component present in the flows and axiomatically hypothesized in the models. Performance indicators and models used for evaluation may also reflect different positions on the 'coordination' axis (see Figure 4.1).

It is fairly obvious that the very use of a model must imply a relatively high degree of 'coordination' between decision-makers, while the use of indicators need not do so. There is also, however, a positive empirical correlation between the 'coordinability' variable and the interpretive capacity of the system. Underlying this is the fact that, as the decision-makers are part of the system, the acceptance of the modelling activity presupposes some coordination between the decision-makers. (It is for this reason that in modelling social and economic phenomena, where situations with a low degree of coordination are examined, e.g. studies of competitive equilibria, interest is focused on the actors' system, whereas in studies of more coordinated systems, e.g. general equilibria, greater emphasis is given to the 'objective' system.)

We now proceed to some more specific questions involved in the construction of models for evaluation purposes. First we summarize a useful check-list of questions which Wilson (1974) suggests the modeller should address and then comment on each in turn.

1. What is the purpose of the model?
2. Which quantified variables should be represented?
3. Which of these variables can the planner control?
4. How aggregate a view can be taken?
5. How should the concept of time be treated?
6. What theory underpins the model?
7. What techniques are available for constructing the model?
8. What relevant data are available?
9. Which methods do we intend to use for calibrating and testing the model?

As far as the first point is concerned Wilson reminds us to construct models which are the most 'suitable' for our purpose. In the above discussion, where we looked at models and the conceptual view of planning they reflect, we attempted to clarify the question of suitability. The real problem is the difficulty of putting this into practice. To construct a completely new model for a particular problem from scratch is a difficult task (although see Openshaw 1989) and to adapt an existing model means taking into account a whole series of constraints imposed by the availability of data, calibration techniques and so on. Great care also needs to be given to the correct choice of model on theoretical grounds, even though the enormous development of urban modelling over the last twenty years has provided such a wide choice of models that problems such as data availability have become in some ways more problematical than the choice of an appropriate theoretical basis.

When we come to the choice of variables, it is obvious that we must include all those which relate to the aspects we wish to evaluate. It is also important, however, when we are making a comparison between alternatives, that the variables identify the causal mechanisms which produce these alternatives. The identification of the subset of variables over which the planner or decision-maker exercises some degree of control is also essential, as these are the instruments which can be used for effectively guiding the system. In relation to the question of the appropriate level of aggregation, it is important to consider the zoning system applied in the model. Zoning always involves a compromise between the need for a detailed description of the geographical distribution of an economic or social activity and the constraints imposed by the availability of data and the computing process. It is also necessary to take into account two specific requirements of an evaluation: first the ease of interpretation of the results (each decision-maker must have access to adequate information relating to the specific aspects of the system with which he or she is concerned); second, the possibility of re-aggregating the zonal results into 'system-wide' indicators, making it possible to contain the total number of alternatives generated.

Particular attention must be paid to the treatment of time in the model. Even though many aspects of comparison can be made without an explicit reference to time (as exercises in comparative statics) an awareness of the evolutionary trajectory of the system becomes important when the timing of the planning intervention or its possible consequences are important aspects of the evaluation. There are particular problems, however, relating to dynamics. In addition to the intrinsic operative difficulties and the lack of a sufficiently thorough theoretical exploration (indicated by the number of new research studies being carried out in this field; see Bertuglia *et al.* 1987b, 1990b) there are still many problem areas in relation to evaluation. It is not always easy in a dynamic model for example to distinguish between independent control variables and dependent variables describing the effects being evaluated as there are often feedbacks between the former and the latter. In an extreme case we could find that the evaluation process is already incorporated in the model and the outcome of the evaluation is therefore predefined.

The question of which theory and techniques to use has already been discussed above. There is no strong distinction between models and indicators in this respect, although there does seem to be a tendency in modelling to use increasingly sophisticated techniques requiring a large amount of quantitative information. In evaluation processes on the other hand growing emphasis is being given to the treatment of imprecise and incomplete (especially qualitative) information. The contrast is not so much in their 'capacity to understand reality' (efforts are in fact being made to provide as rigorous a treatment as possible of the qualitative aspects) but where they choose to place the emphasis – on the intrinsically qualitative or on the more easily measurable aspects of the system. A useful area for future research would be to find a more organic way of connecting the two.

Next we come to the question of data availability. Although it is true that evaluation techniques can be applied even when some data are missing, it is nevertheless preferable to have as much data as possible, even if they have been 'reconstructed' using a model. Data reconstruction is one of the roles played by the simulation model.

Having discussed these broad ideas concerning evaluation, models and performance indicators it is now useful to begin to articulate the types of indicators we are clearly interested in.

4.6 PERFORMANCE INDICATORS FOR EVALUATION

To transform raw data into meaningful information it is necessary first to decide the end purpose, i.e. to identify its logical relationship with the objectives so that a given piece of data can become for example an effectiveness or efficiency indicator. Second, it is important to locate it in a logical scale of measurement with respect to its purpose and then to relate it to other data in order to compact the information, eliminating redundant data and making it congruous with the functional interdependences of the system.

This process normally means that in practice the data have to be manipulated or combined with other raw data to arrive at the 'derived' data which is the performance indicator. Performance indicators can be calculated either from the model input (or some other data) or its output. It may be that having decided to calculate the indicators directly from some specific data we find that the necessary information is incomplete or not available. In this case the missing data can be calculated using mathematical models.

It is useful at the outset to think in terms of a threefold classification of indicators, ranging from the 'descriptive' indicators seen in Chapter 2 to more complex 'profile' and 'performance' indicators. To recap, descriptive indicators would include information based on simple algebraic operations such as sums, percentages and products and statistical variables such as means and variances. Profile indicators can be thought of as ratios, such as the degree of spatial homogeneity in, for example, population types, housing types and job types or the degree of 'self-containment' within small areas such as inward and outward commuting flows. The third kind of indicator, we argue, must be able to reflect the systemic interdependences in urban systems and hence be based on theories of spatial interaction. We shall introduce the ideas on such performance indicators here, leaving the detail until Chapter 5.

Urban systems consist of a population, a complex set of economic sectors and the interactions between the two. The economic sectors can be structured into a number of networks or 'organizations' (e.g. the retail trade network) which provide services for, or are in some way utilized by, the population. The population too can be structured into organizations, e.g. the household, but this is less relevant to the present discussion.

There are then two main kinds of performance indicator. The first provide a way of measuring the extent to which the population is served by, or utilizes, the service organizations referred to above. These are the *effectiveness indicators* and they are calculated for each zone of residence of the population. The second measure the extent to which the organizations are utilized by the population, i.e. they are *efficiency indicators* which are calculated for the zone in which the organization is located. These indicators differ from the classical social indicators in that they are able to take into account spatial interaction between the population and organizations. They should not necessarily be considered as alternatives, however, but additionally as a useful complement.

The efficiency and effectiveness indicators are based on the concepts of 'potential service provision in a zone of residence' and 'catchment population of a service located in a zone'. To clarify this idea let us imagine an urban service such as retailing. By 'potential service provision in a zone of residence' we mean the availability of shops in the various zones as it is 'perceived' by the residents. The catchment population refers to the total resident population making use of the shops. These are by no means the only kinds of performance indicator useful in the evaluation process. There are many others which can make a valuable contribution. We shall expand on these arguments in the next three chapters. Such indicators can be derived from either subsystem models (labour markets, incomes, retailing, education, health etc.) or from the outputs of more comprehensive models. In the next chapter we look at these subsystem models in greater detail, with applications shown in Chapter 8. Chapter 9 deals with applied indicators based on comprehensive models and we conclude this section with a recap on the importance of what we have continually referred to as 'second-generation models'.

In the construction of model-based performance indicators, both for reasons of logical coherence and in the interest of obtaining the best use from the model, we need to be aware of the 'vision of the city' that the specific model represents. Frequently the model itself helps us define performance indicators which are appropriate to the needs and objectives of our society, in the sense that they emerge from the understanding provided through the model of the socio-economic and geographical systems which make up the city.

A Lowry-type model is therefore still a good base for deriving indicators associated with the functional relations of the urban system, especially in relation to the spatial distribution of the functions and those connected with physical movement of people and goods. Good examples are the accessibility indicators and 'systemic' indicators of spatial equity and efficiency that we have seen in Chapter 3 and above.

The second generation models, however, are more suitable for constructing the following kinds of indicators: first, those focusing on significant economic and social features of modern society (the interdependences in the urban system instead of being analysed in terms of the usual economic categories can be analysed in terms of other characteristics, e.g. different racial components of the population, levels of education etc.); second, those indicators relating to the modern forms of production of goods and services, e.g. the location of manufacturing activities can take into account the growing interconnectedness ('network' structure) of the economy and the development of advanced tertiary services; third, those which measure the effects induced by the new technologies (in this context it is necessary to consider above all the change in location patterns and spatial interdependences resulting from the new forms of telecommunications); fourth, those which reflect new dimensions of urban problems such as environmental issues – the question of pollution for example should be included as an essential component of urban quality and as a factor of interdependence between industry and housing (in the case of industry which is dangerous or produces pollution); finally, we are interested in those which require an adequate treatment of the time dimension – this is necessary not only when we are dealing with the evolution of the system but also when we wish to focus on sudden unexpected transitions, elements of irreversibility, the inertia of change and how this affects the existing situation.

Above all, the second-generation models reflect the difficulty in socio-economic systems of making a clear separation between the system of decision-makers and the subsystem being 'governed'. This leads us to the recognition that the process of analysis and the project can no longer be considered in watertight compartments but are fundamentally interconnected (Rabino forthcoming). A comprehensive set of indicators must therefore be able to reflect the essential interrelatedness between the two subsystems. It should include indicators which are able to capture the degree of conflict between the decision-makers and also the extent to which the decision-makers involved can exercise 'control' over the system (providing a measure of the variables which they are effectively in a position to govern) and hence the existing scope for operating effective planning policies.

4.7 THE SCIENTIFIC METHOD IN URBAN PLANNING

We have described how mathematically based methods came to be introduced into the planning process and used with some success for analysis and as decision-support tools. It is time to conclude with a reassessment of the role of models (and more generally the scientific method) in planning and evaluation. The systems approach not only has influenced the way in which the urban system is conceived but has also raised fundamental questions about what 'planning' is trying to achieve and how it should go about it . For this reason it is worth examining the conceptual premises which underlie the application of a mathematical model. (We should specify here that we are using the term mathematical model in its widest sense, including not only simulation models but any other formalized method including all the various statistical techniques or optimizing algorithms. The application of mathematical models implies the adoption, and therefore acceptance, of the scientific method.)

According to Hay (1985) the scientific method can be defined as a way of thinking in which four ingredients come into play: theory, regularity, logic and reduction (see also Bertuglia and Rabino 1990). The scientific method is an extremely powerful instrument which has been effectively applied in many disciplines. To maintain that this is the *only* way of acquiring knowledge, however (an attitude we can define as *scientism*), can be dangerous for two reasons:

1 the temptation to apply the scientific method in contexts which are not appropriate;
2 the risk of ignoring issues which cannot be investigated with the scientific method.

There are other reasons why scientism must be considered unacceptable. Studies of the philosophy of science point out that in the course of history many of the important scientific breakthroughs resulted from logical 'jumps' or 'flashes of intuition' and not only from the application of a rigid process of 'scientific' reasoning. The difficulties of applying the scientific method satisfactorily become increasingly evident when we are dealing with disciplines such as the social sciences and urban planning (or regional science). This kind of application raises a number of serious problems, some of which we shall now discuss.

The first concerns the question of the uniqueness of the systems involved. Urban systems are widely recognized as being highly complex, because of the large number of facets they possess and the complicated causal relationships between them. This is due largely to the specific space–time characteristics of the systems. We cannot altogether exclude the hypothesis that if these characteristics are totally specific to a particular space and time then the system must in the extreme case be considered totally unique, making the formulation of any generalized theories impossible.

A second problem, also connected with the complexity of this kind of system, is the difficulty of carrying out any real scientific experimentation. Laboratory testing is obviously impossible and field experiments, given the complexity of the systems, run into the obstacle of the control of the surrounding conditions (e.g. the possible influence of outside variables). Another particular aspect of this problem concerns the interference of the 'observer' with the 'observed' phenomenon. In fact, where interaction with 'thinking' beings is involved, this may create alterations in the behaviour of the system which in extreme cases can be irreversible.

A third problem is whether, and how, we should transfer theories or laws from other disciplines to urban and regional science. A transfer could be considered valid if the urban 'sciences' are thought of as a unification of other more elementary sciences, or 'spatialization' is seen as a specific case of a more general science. If we think for a moment, however, of the number of disciplines which touch on planning issues or the many ways in which interactions may occur, it is clear that the problem of how to reconcile them all is enormous, and at a certain point may even become impossible.

What opportunities remain then for a valid application of the scientific method, and therefore of mathematical models in urban and regional planning? Our own view is that while the various problems outlined above should clearly be recognized and outright scientism should be avoided, there are sound reasons for not rejecting the scientific method out of hand. First, the dangers outlined above represent extreme cases. There remains a broad area where the scientific method can provide a fruitful way of constructing and testing theories to explain urban phenomena. In addition, the scientific method continues to be attractive in as much as it codifies typical everyday thought-structures, including the desire for theories which we can test from experience (we establish whether they are 'correct' through a trial and error procedure, i.e. an interaction between theoretical formulation and practical experience). Finally, as a consequence of the rapid spread of new technological developments and in particular the use of computers, knowledge expressed in a 'scientific' or mathematical form is increasingly requested by society for the management of its systems.

This position should not be misunderstood as an unqualified defence of existing models and experimental procedures. It is possible, and also desirable, that an improvement of the scientific tools used in urban planning be achieved and interest be stimulated in the furthering of studies relating to the philosophy of urban science. This would help pinpoint the sources of shortcomings in current practice and indicate the most productive directions for future research. The growing interest already evident in methods for the rigorous treatment of non-quantitative information (the development for example of another abstract language suited to the logic of social and urban phenomena, such as multi-criteria methods) is a sign that something has already been achieved in this direction.

Many of the questions relating to the pertinence of the use of the scientific method in urban planning are due, as we have seen, to the sheer complexity of urban and regional systems. This is not the moment to go further into the concept of complexity (the reader can find a more detailed discussion of this question in Waddington (1977) for example). It is sufficient to observe that the image of reality provided by science is intrinsically hierarchical. The systems identified at a given level of observation are in general broken down into a number of subsystems which are systems of a lower level of complexity. We therefore have a hierarchical scale of complexity in which the urban systems can be considered to occupy a high rank, since they are more complex than the economic, social, environmental and other systems which go to make them up.

Since the seventeenth century Western science has tackled the problem of the comprehension of systems in two different ways. The first is an analytical approach developed in the more elementary sciences, i.e. those, such as physics and chemistry, situated on the lower 'rungs' of the ladder of complexity where it served its purpose admirably (so well that it came to be considered as practically the only method). According to this approach the characteristics of a system of a given level of complexity can only be fully understood by breaking it down into component parts, i.e. tackling the problem at a lower level in the hierarchy of complexity (hence the properties of molecules are explained in terms of atoms, cell biology in terms of their chemical constituents, and so on). This we term *reductionism*.

The second approach which is considered by some to belong to the scientific method, by others to fall outside, suggests that phenomena of a given level of complexity should be approached on the same level of complexity rather than a lower one. This means that an attempt must be made to understand the whole rather than the constituent parts and that the whole is *more* than the mere sum of the parts, i.e. that it has specific properties which cannot be predicted only from the knowledge of its parts. This approach is termed *holism*. (The uniqueness of each human being provides a good example of this idea.)

Reductionism has in the past been widely applied to urban and regional studies, the identification of 'laws' at a given level of the system being explained by consideration of the behaviour of lower-level elements, as for example in Wilson's entropic derivation of spatial interactions (1970). But it is only fair to add that the literature also contains many examples of exceptions to these laws and recognition of the limitations of the results obtained.

With the introduction of the concept of the *ecosystem* (a dynamic interactive system linking all living organisms and inorganic material in a given area and at all scales) holism began to find its place in the analysis of urban phenomena. A typical description of an ecosystem is in terms of cycles of materials and flow of energy through the system. In recent years the view of urban and regional systems as 'human ecosystems' has become increasingly popular. This has occurred partly as a result of the growing attention given to morphogenic phenomena and the irreversible and self-organizing processes recognized in these systems (Wilson 1981a; Allen 1982) which can only be analysed by treating the system *in toto*.

Whether holism is destined to predominate over reductionism is difficult to say. Certainly the holistic approach has helped to focus attention on certain important issues such as the impact of the population explosion, the technological revolution and environmental pollution. The development, however, of corresponding theories and experimentation, such as that attempted by Allen, is still at an early stage and openly admits to the limitations of the results obtained (in a sense not unlike the reservations encountered in reductionist-type studies). Bearing in mind that the choice between the two approaches is not only conceptual but also has operative implications, what position should one adopt? We are in agreement with Wilson (1981b) that urban and regional science should continue to be considered a discipline of synthesis, but that a concerted research effort should be made to construct efficient new theories of systems *in toto*. In other words, the aim should be essentially holistic, even though the methods applied may remain for the moment essentially reductionist.

As stated previously, the choice of approach also has operative implications. This can be illustrated by the process of calibration of an urban model. Lombardo *et al.* (1987) point out that if we adopt a reductionist approach the model can be broken down into submodels (assuming that these are relatively independent) and the spatial parameters of these submodels can then be correctly calibrated at given moments in time (assuming that the subsystem is relatively close to equilibrium at the moment considered). If, on the other hand, the approach is holistic these two hypotheses no longer hold and the model parameters can *only* be calibrated jointly from data relating to the time trajectory of the whole system.

This second alternative, as it avoids approximations, would appear in theory at least to be preferable, but in reality the problems of statistical inference in the calibration of the model taken *in toto* are far more complex than those of the single submodels. Consequently it is difficult to say whether or not the resulting increase in realism, but also complexity, is preferable to the acceptance of certain approximations. For the moment at least a practical proposition would appear to be to adopt the reductionist approach for operative applications and make the problem of the reduction of approximations a priority for theoretical research.

5 A new geography of performance indicators for urban planning

G.P. Clarke and A.G. Wilson

5.1 INTRODUCTION

The argument of Chapter 2 was that most existing indicators are calculated in relation to *data*, and in relation to single zones with little reference to other zones. In addition, many of the indicators are not directly related to what we want to measure – 'quality of life' in some way. A key argument of the previous chapter was that a model-based approach can be used to make us less reliant on data availability – model predictions can be used to fill 'gaps'. It can be focused to zonal properties which depend crucially on interdependence with other zones – job opportunities, availability of services, and so on. It can be focused directly on people and organizations rather than 'areas' *per se*. In addition, the indicators are calculated from variables which are systemically related and are directly connected to our knowledge of urban structure and processes. The task now is to show in more detail what is involved in implementing them.

5.2 KEY CONCEPTS: EFFECTIVENESS, EFFICIENCY AND SYSTEMIC INTERDEPENDENCE

The basic position which is taken at the outset is that cities are made up of their residents and a complex set of organizations which are created by, and which can be considered to serve, these residents. The physical structures should be related to the processes associated with individual, household and organizational activities. Performance indicators should therefore relate to the activities of households and organizations and the way in which they relate to each other. The following discussion builds on the distinction drawn in Chapter 4 between residential and facility-based indicators.

There are two fundamentally different kinds of performance indicators: those which relate to individuals and households, rooted in residential location but related to the ways in which they are served by organizations; and those which relate to the efficiency and roles of different kinds of organizations (again in relation to their location). To an extent, of course, some organizations should be evaluated in relation to the ways in which they serve other organizations, but that is a complication which we postpone until later.

The argument is based on the notion of *spatial interaction* and in this sense it differs from the traditional approach to the definition of indicators seen in Chapter 2. Nearly all indicators in past analyses have been defined in relation to individual spatial zones; but they have not been connected to other zones. On this basis, it is possible to identify deprivation in residential areas but not to address the issue – for instance – of how levels of deprivation may be related to different types of individual positions in a spatially-wider labour market. In this section, we attempt to provide the concepts which overcome these problems.

To fix ideas, let us focus on the provision and use of some service – shops, schools, hospital beds or whatever. Assume that an aggregate portrait is adequate to help us introduce the basic concepts; and so let e_i be the per capita demand for the service in zone i (possibly elastic) in relation to a population P_i, and let W_j be the scale of provision in j. Let S_{ij} be a measure of usage of the facilities in j by residents in i. Let c_{ij} be a measure of cost of travel.

There are now two fundamentally different kinds of indicator: those which relate to individuals and households at the residential location; and those relating to organizations at the facility location. These can be thought of as distinguishing *effectiveness* of delivery at residential locations from *efficiency* of provision at facility locations. It will emerge that it is possible to be effective without being efficient; or vice versa; or, of course, it is possible to achieve neither (see Chapter 2, Section 2.5). The notion of spatial interaction can be used to link these two kinds of indicators, and it is this linkage which provides the basis for understanding systemic interdependence.

To make progress, we have to (re)introduce two concepts seen in the previous chapter: 'effective delivery' at a residential location, and 'catchment population' at a facility location. These can be constructed in relation to the example as follows. Define

$$\hat{W}_i = \sum_j \frac{S_{ij}}{S_{*j}} W_j \tag{5.1}$$

$$\Pi_j = \sum_i \frac{S_{ij}}{S_{i*}} P_i \tag{5.2}$$

In equation (5.1), a typical term on the right-hand side involves taking the *proportion* of provision at j which is used by residents of i and then summing to obtain a measure of total provision for residents of i. Hence this can be called 'effective delivery'. Similarly, the terms of (5.2) represent partitions of residential population which are then combined to form a catchment population for the 'centre' (or outlet) at j. It may also be useful to calculate a version of this in terms of demand, which we can call Δ_j:

$$\Delta_j = \sum_i \frac{S_{ij}}{S_{i*}} e_i P_i \tag{5.3}$$

Models are needed to predict S_{ij}. If a standard spatial interaction model is used, then (5.3) reduces to 'revenue' (if the service is marketed) since

$$S_{ij} = \frac{e_i P_i W_j^{\alpha} \exp(-\beta c_{ij})}{\Sigma_k W_k^{\alpha} \exp(-\beta c_{ij})} \tag{5.4}$$

and this implies

$$S_{i*} = \sum_j S_{ij} = e_i P_i \tag{5.5}$$

Typical indicators are then $\hat{W}_i/P_i$ for effectiveness, and W_j/Π_j for efficiency. What is virtually meaningless in many situations is the traditional indicator, which would be taken as W_i/P_i or W_j/P_j.

These indicators become much more interesting when the population is disaggregated by type *m* (say social class, car owner/non-car owner) and provision is disaggregated by type of good (g). Indeed the definition of the variable P_i^m is likely to be crucial in the equity–efficiency issue. One of the problems of an indicator-based study highlighted in Chapter 2 was the so-called 'ecological fallacy' of attributing average conditions to the entire population of the area under study. Whilst the focus on homogeneous populations may be useful in exposing general levels of welfare or deprivation, it is important to appreciate that these will vary between different sections of the community. Indeed, most of the urban problems of today could be classified in terms of gender, class and race and the often unique problems each of these groups face. The Women and Geography Study Group (1984), for example, explore the restrictions on access to facilities by women, especially those caring for young children who have specific problems of mobility and hence demands on local facilities.

We have thus distinguished two very different aspects of performance indicators in cities. This is achieved by focusing on interaction and then distinguishing catchment population from residential population to provide a device to assess performance in relation to production as something different from performance in relation to delivery.

The obvious planning task within a service sector g is to ensure that the spatial distributions of provision and expenditure, $\{W_j^g\}$ and $\{C_j^g\}$ are such that *both* sets of effectiveness and efficiency indicators achieve appropriate targets. (In many situations it is possible to achieve neither; but in many others, one rather than the other.)

In the case of marketed goods and services, it is possible to use, in addition, revenue attracted to g-type organizations in j as part of performance assessment. Indicators like

$$\frac{D_j^g}{W_j^g} \qquad \frac{D_j^g}{C_j^g} \qquad D_j^g - C_j^g$$

will then be appropriate. In this case, the ability of individuals to purchase will obviously be dependent on income; and, in the main, incomes will be dependent on another function of the city in relation to its people – as a supplier of jobs.

In all these cases, it may be important to distinguish in some way the *quality* associated with residential environments and the goods and services provided. We also need to distinguish the transport costs (in some 'generalized' sense) incurred by individuals in 'using' the organizations of the city. We can then build up a broad picture of the set of performance indicators to which this argument leads. They are summarized in Table 5.1.

We have seen that, usually, it will be important to base the definition of many of the indicators of both types on the actual levels of *interaction* between households and organizations; and this fundamentally changes the usual approach which is more likely to be based on residential zones alone. This argument therefore establishes the principles on which we can proceed and we now make some recommendations for a framework which is consistent with these in the next section; we compare the result with the illustrations we presented of alternative frameworks and ideas in Chapter 2.

5.3 RESIDENCE-BASED INDICATORS

Let us look at each of the headings in Table 5.1 in more detail. In each case, we shall look at traditional indicators used in planning and a new set of indicators derived from our model-based framework.

5.3.1 Income

Income has prime importance because, if it is sufficiently high, then it is easier

Table 5.1 'Sets' of performance indicators

Residence-based
Household incomes
Quality of housing and residential environment
Work opportunities
Take up of marketed goods and services
Take up of public goods and services
Transport (generalized) costs
Organization-based
Efficiency of production
Role in the pattern of provision
Role in the labour market

to assume that the perceived outcomes of choices by individuals and households are less constrained, i.e. are more likely to represent a desired situation. Then, the basic dimensions of life can be related to housing, work and a wide range of services. Because incomes are not directly available from data in the UK, traditional indicators have been based on car ownership, socio-economic group and so on. (The United States has the opposite problem – how to define social group classifications from income data.)

It is essential for effective planning to be able to identify the sources of income: employment, social security or as returns from different kinds of wealth. Conventional economic analysis is not usually helpful at this point and the kind of analysis offered here should turn out to be a valuable alternative. First, it is obviously necessary to distinguish employed from unemployed in the population of economically-active individuals; then, individuals of all kinds will have to be combined into households, since we need a knowledge of both person and household incomes, and the ways these are combined in households. If we define w as person type (age, sex, race etc.) b as occupation type and g as industry type then crucial arrays are P_i^{wb} and E_j^{gb} and the labour market resolution, say T_{ij}^{wgb}. The latter array refers, obviously, to those who actually find a job. It is also necessary to calculate (say) a probability p_i^{wb} that someone in the (w, b) group in zone i actually has a job. (There are difficult issues of dynamics here which can be handled but will be neglected for simplicity for the time being – for instance that the probability of living in i at a particular time will be dependent on previous employment history.) Then if $\hat{I}_j^{gb}$ is the average annual wage or salary of the (g, b) groups in j, then $T_{ij}^{wgb}\hat{I}_j^{gb}$ is the amount taken back to i by (w, b, j, g) groups. By computing unemployment from P_i^{wb} and p_i^{wb} and relating (w, b) potential workers to their household structures, it should also be possible to calculate social security income (as well as income derived from 'wealth'). The outcome of a complicated submodel would be an estimate of (say) I_i^h, the average annual income of a type-h household in zone i. If aggregate *area* indicators were required, they could be computed; but it is much more effective to have the array $\{I_i^h\}$ together with the $\{T_{ij}^{wgb}\}$ elements which provide a rich source of additional information on where the different kinds of income derive from. However, it should be recognized that the emphasis on income is a new focus for this style of urban modelling. A variation on this approach, using micro-simulation techniques, is described in Birkin and Clarke (1989). We shall explore income generation in more detail in Chapter 8.

5.3.2 Housing and quality of residential environment

The issues in housing and quality of residential environment are also more complicated than is usually reflected in urban modelling practice. We need to be concerned with the physical characteristics of housing, but also with an index of the quality of maintenance: there are many instances of what were originally 'grand' houses being run down (e.g. in decaying areas) and used by different kinds of households. And, indeed, vice versa: 'workers' houses in what have become fashionable areas are upgraded. These examples also show the connection between housing quality and the broader 'area' notion of 'quality of residential environment' – which is itself partly a physical feature and partly socio-economic. These gradations of housing types, whose physical characteristics are otherwise the same, can in part be captured by price.

In practice, we are going to have to define a vector of housing characteristics, ***I***, which will be partly physical and partly environmental and socio-economic. At one level, it will be possible to define *standards* below which quality is unacceptable. But the argument is implying that we should also be trying to be more ambitious.

Ideally, we need a measure of the utility of living in a house of a particular type (with or without a measure of the locational advantages included). In some ways, a composite measure might be taken as observed 'price', but there are difficulties with this. First, such prices may reflect shortage of particular house types in a particular study area. Second, they may not reflect accurately what people pay, which will be a function of past loan history, wealth and so on. We return to these issues in relation to composite indices below. To make progress let us define H_i^k as the number of type-k houses in i, r_i^k as the notional annual 'rent' and T_i^{hk} as the number of type-h households resident in type-k houses in i. This array is a rich source of information. If we could define k_c as some cut-off category in k below which the housing standard was unsatisfactory, then it would also be possible to calculate aggregate zonal indicators such as

$$\sum_{h,k<k_c} T_i^{hk} \Big/ \sum_{hk} T_i^{hk} \tag{5.6}$$

as the proportion of households in i living in substandard housing. It could be coupled with T_{ij}^{wgb} arrays (Section 5.3.1) so that the sources of the income which supported different types of resident in i could be investigated.

5.3.3 Work opportunities

The main arrays involved have been discussed above in relation to incomes. If we separate age (a) from our category of other person types (w) then we increase the size of our main array to $\{T_{ij}^{wbag}\}$. The labour force available in i is P_i^{*ba*}. We can use the argument of Section 5.2 to derive a measure of job 'availability' for occupation group b in i as follows:

$$\hat{E}_i^{ba} = \sum_{jg} \frac{T_{ij}^{*bag}}{T_{*j}^{*bag}} E_j^{*bag} \tag{5.7}$$

These values are reflections of the workings of the labour market as recorded in $\{T_{ij}^{wbag}\}$. An indicator such as

$$p_i^{ba} = \frac{\hat{E}_i^{ba}}{P_i^{*ba*}} \tag{5.8}$$

then provides a measure of employment effectiveness for the residents of i. p_i^{ba} can be interpreted as the probability of an individual in the (b, a) category, resident in i, obtaining a job. Conversely, $1 - p_i^{ba}$ is the probability of being unemployed. This kind of indicator is obviously of the greatest importance. There is the possibility of attempting to understand the spatial variations in $\{p_i^{ba}\}$ in terms of a model-based calculation of $\{T_{ij}^{wbag}\}$.

5.3.4 Marketed goods and services

It might at first be thought that goods and services which are provided through a market will automatically be delivered effectively. But this is not necessarily the case: the *location* of such facilities has the characteristic of a *public* good. If you happen to live near to a facility, then the additional benefits of this are available free of charge (unless they are to some extent reflected in house prices), and vice versa. It is therefore valuable to analyse provision with this in mind.

The usual indicators are shops (possibly by type) per head of population (which are grossly inadequate in relation to interaction) and accessibilities (which are better but still not highly informative). As is usual by now, we can improve the situation by constructing new indicators based on interaction.

We can use the argument of Section 5.2 directly, but again it is useful to disaggregate: let the interaction array be S_{ij}^{mg}; let e_i^{mg} be the per capita demand for a resident population, P_i^m; let W_j^g be the scale of provision at j for g; let c_{ij} be the transport cost as usual; m is an aggregation of person types (w, b). Then we can calculate effective provision at i for m as

$$\hat{W}_i^{mg} = \sum_j \frac{S_{ij}^{mg}}{S_{*j}^{mg}} W_j^g \tag{5.9}$$

and, in aggregate,

$$\hat{W}_i^{*g} = \sum_m \hat{W}_i^{mg} \tag{5.10}$$

The catchment population at j is

$$\Pi_j^g = \sum_{im} \frac{S_{ij}^{mg}}{S_{i*}^{m*g}} P_i^{*m} \tag{5.11}$$

Then indicators of effectiveness of delivery are

$$v_i^{mg} = \frac{\hat{W}_i^{mg}}{P_i^m} \tag{5.12}$$

and of efficiency

$$\hat{v}_i^g = \frac{D_j^g}{\Pi_j^g} \tag{5.13}$$

where

$$D_j^g = \sum_{im} S_{ij}^{mg} \tag{5.14}$$

is a measure of total usage or total revenues.

In this case, it is useful to investigate the consequences of representing the flows in a conventional model. Suppose that

$$S_{ij}^{mg} = A_i^{mg} e_i^{mg} p_i^m (W_j^g)^{\alpha^m} \exp(-\beta^m c_{ij}) \tag{5.15}$$

where

$$A_i^{mg} = \frac{1}{\Sigma_k (W_k^g)^{\alpha^m} \exp(-\beta^m c_{ij})} \tag{5.16}$$

to ensure that

$$\sum_j S_{ij}^{mg} = e_i^{mg} P_i^m \tag{5.17}$$

Inspection shows that there is no dramatic simplification if we substitute for S_{ij}^{mg} from (5.15) or S_{i*}^{mg} from (5.17) into the earlier equations. There is one new feature, however. It has been shown (in Chapter 3) that

$$b_{ij}^{mg} = \frac{\alpha^m}{\beta^m} \log W_j^g - c_{ij} \tag{5.18}$$

can be taken as a measure of consumer surplus for the (i, m) group using (j, g) facilities. An appropriate aggregate measure is therefore

$$B_i^{mg} = \sum_j b_{ij}^{mg} \tag{5.19}$$

and

$$B_i^{mg} / P_i^m \tag{5.20}$$

should be an interesting indicator.

Appropriate residential zone indicators are then ratios like

$$\frac{\hat{W}_i^{mg}}{P_i^m} \quad \frac{\hat{W}_i^{*g}}{P_i^*} \quad \frac{\Sigma_{ij}\ b_{ij}^{mg}}{P_i^m} \quad \frac{\Sigma_{jm}\ b_{ij}^{mg}}{P_i^*} \tag{5.21}$$

It may also be appropriate to examine the total transport costs by m-residents of i for g,

$$C_i^{mg} = \sum_j S_{ij}^{mg}\, c_{ij} \tag{5.22}$$

but this is one component of total benefit. Finally note that we can calculate

$$\bar{c}_i^{mg} = \frac{\Sigma_j\ S_{ij}^{mg}\ c_{ij}}{\Sigma_j\ S_{ij}^{mg}} \tag{5.23}$$

as the average travel cost of an (i, m) resident for g – though recall the earlier comment that such an indicator should be interpreted carefully within a composite index because, for example, individuals may be trading high travel costs against other benefits.

The above argument has been carried through on the basis that most services are used in relation to residential location. In some cases, there is considerable use from, say, workplaces. In other cases, say for garages and sales of petrol, the spatial relationships may be even more complicated. Appropriate modifications may be required, and there is extensive literature on alternative attractiveness and deterrence terms (Fotheringham and O'Kelly 1988). In the case of efficiency indicators, the modification is simply a matter of adding in the additional flows to the calculation of revenues. In the case of attractiveness, in cases where the flows are significant, new indicators would have to be defined measuring the effectiveness of (say) delivery to workplaces.

Although we have wide experience in designing suites of performance indicators for retail analysis we have deliberately kept the arguments brief here. Birkin, in Chapter 7, builds on the arguments to define an applied set of performance indicators for retailing (some of which are shown 'in use' in Chapter 8).

5.3.5 Public goods and services

The same basic argument applies to public goods and services as for marketed goods and services except that the units will not usually be financial. That is, per capita demands e_i^{mg} will be measured in terms of (say) trips. In many cases it may also be useful to calculate indicators based on different aspects of provision. That is, the W_j^g terms may be measured in different units within the indicator calculations. In the case of health services, for example, it may be appropriate to measure W_j^g in terms of beds and to use this in formulae such as (5.9). But then C_j^g can be taken as expenditure and this may be taken as an indication of quality of service (provided efficiency can be guaranteed!), and we could calculate

$$\hat{C}_i^{mg} = \sum_j \frac{S_{ij}^{mg}}{S_{*j}^{mg}} C_j^g \tag{5.24}$$

with the corresponding indicator

$$\hat{C}_i^{mg} / P_i^m \tag{5.25}$$

as notional expenditure per capita within group m. This begins to provide a direct measure in suitable cases of the way in which public expenditure is allocated to particular groups (by location).

We have kept this section deliberately brief since changing political attitudes to the welfare state in general in the UK has meant that public services have been subjected to fundamental reform. The changing circumstances of public sector service delivery warrant greater consideration of the historical role of performance measurement and call for new indicators based partly on those necessary for marketed goods and services. We shall explore both residence-based and organization-based indicators for health and education together in Section 5.5.

5.3.6 Transport

Each sector above has an analysis based on models of spatial interaction. Such flows are, of course, transport flows. However, it may also be appropriate to calculate transport flows directly. In effect, this means collecting the flows together on a different time basis (a 'peak period' or a day, say) and calculating arrays like T_{ij}^{mkp}, where m is person type as before, p is trip purpose and k is now emphasized as transport mode. Because these person trips are assigned to a transport network, it is possible to obtain estimates of modal costs c_{ij}^k which reflect congestion. Ideally, such costs should be *generalized* costs of the following form:

$$c_{ij}^k = m_{ij}^k + at_{ij}^k + be_{ij}^k + p_i^{(1)k} + p_j^{(2)k} \tag{5.26}$$

where m_{ij}^k is the money cost, t_{ij}^k the travel time, e_{ij}^k the 'excess' time (waiting at bus stops etc.) and $p_i^{(1)k}$ and $p_j^{(2)k}$ are the origin and destination costs respectively. It is the t_{ij}^k term which reflects congestion, of course; the other terms are part of the system description.

The c_{ij}s which have been used in the earlier analyses can be thought of as some kind of average of the c_{ij}^ks. It has been argued (e.g. Wilson 1970) that exponential averaging is best:

$$\exp(-\beta c_{ij}) = K \sum_k \exp(-\beta^{-k} c_{ij}) \tag{5.27}$$

for a suitable constant K and parameter β. They provide the crucial connection between the transport model and the other models within a comprehensive system.

Changes in transport policy – e.g. on network development – will influence the c_{ij}s and, through them, many of the indicators defined earlier. In this way, it is possible to use transport policy directly, and it is to these which we now turn.

We can calculate total transport 'costs' (in generalized cost units) for m-type residents of i for purpose p as

$$C_i^{(T)mp} = \sum_{kj} T_{ij}^{mkp} c_{ij}^k \tag{5.28}$$

and then a suitable per capita indicator is

$$C_i^{(T)mp} / P_i^m \tag{5.29}$$

(where P_i^m refers to population total). These quantities could also be calculated for each term of the generalized cost:

$$C_i^{(1)mp} = \sum_{kj} T_{ij}^{mkp} m_{ij}^k \tag{5.30}$$

$$C_i^{(2)mp} = \sum_{kj} T_{ij}^{mkp} t_{ij}^k \tag{5.31}$$

$$C_i^{(3)mp} = \sum_{kj} T_i^{mkp} e_{ij}^k \tag{5.32}$$

$$C_i^{(4)mp} = \sum_{kj} T_{ij}^{mkp} p_i^{(1)k} \tag{5.33}$$

$$C_i^{(5)mp} = \sum_{kj} T_{ij}^{mkp} p_j^{(2)k} \tag{5.34}$$

The corresponding per capita indicators can be computed along the lines of (5.29). Equation (5.34) might be more usefully computed for destination zones:

$$C_j^{(6)mp} = \sum_{ki} T_{ij}^{mkp} p_j^{(2)k} \tag{5.35}$$

In order to get a per capita version of this, we need to calculate catchment populations by trip purpose – and the pattern of these may be useful indicators in themselves:

$$\Pi_j^{mp} = \sum_{ik} \frac{T_{ij}^{mk}}{T_*^{mkp}} P_i^m \tag{5.36}$$

with indicator

$$C_j^{(6)mp} / \Pi_j^{mp} \tag{5.37}$$

However, from the point of view of the *particular traveller*, the components of c_{ij}^k are the unit costs directly: the indicators represent a kind of apportionment.

A popular indicator which relates partly to transport provision is well known and turns on the idea of accessibility (cf. Chapter 2). This can be usefully added to our list here:

$$X_i^{mp} = \sum_{kj} D_j^p \exp(-\beta^{mp})\, c_{ij}^k \tag{5.38}$$

where $\{D_j^p\}$ is the array of destination opportunities for purpose p. Parameters β^{mp}, assumed to be taken from a model, now have to be introduced for the first time.

Another indicator which is less directly related to accessibility is important if trip generation is elastic. This is based on residential zones:

$$O_i^{mp} \,/\, P_i^m \tag{5.39}$$

where O_i^{mp} is the total number of trips generated in i for purpose p by type-m people. This represents something like 'proportion of need' met.

It is useful to calculate an economic indicator in the form of consumer surplus. This (as with accessibilities) assumes an underpinning model and is best calculated in terms of change from t to $t+1$, and so time labels are added to the appropriate variables (cf. Williams 1977):

$$\Delta B(t, t+1) = \sum_{kmp} \frac{1}{\beta^{mp}} \log \frac{\Sigma_{ij} \exp[-\beta^{mp}(t,\ t+1)]\ c_{ij}^k(t,\ t+1)}{\Sigma_{ij} \exp[-\beta^{mp}(t)]\ c_{ij}^k(t)} \tag{5.40}$$

If the summation on the right-hand side is restricted to j, then an estimate of change in consumer surplus for m-residents of i can be obtained, and the corresponding per capita index is

$$\frac{\Delta B_i^m(t,\ t+1)}{P_i^m(t)} \tag{5.41}$$

If total capital expenditure in the period is $\Gamma(t,\ t+1)$, then we might apportion this to residents in proportion to ΔB_i^m, and then construct an appropriate per capita index:

$$\frac{\Delta B_i^m(t,\ t+1)}{\Delta B(t,\ t+1)} \frac{\Gamma(t,\ t+1)}{P_i^m(t)} \tag{5.42}$$

Transport policy can be expressed in relation to capital investment or perhaps in relation to changes in particular components of generalized cost. The indicators constructed above can be used to assess the direct impacts of such policies. The deeper (or 'secondary') effects can be traced through the impact of $\{c_{ij}\}$ changes on all other indicators. Some of these effects might be quite subtle. For instance, c_{ij} changes will bring about changes in catchment populations, which will change the denominator of various per capita indices.

5.3.7 Composite indices

We have already emphasized that it is necessary to be able to look at the ways in which people trade off the benefits or costs of one kind of indicator against others. We now try to be more explicit about this and the construction of an argument forces us to be more explicit about some features of particular indicators.

Suppose we focus on a type-m individual or a type-h household resident in zone i. We consider first the individual and then the household – and the distinction generates its own problems which have to be solved. To simplify in order to proceed with an exploration, it may be best to think that we are dealing with a type-m individual who is the *only* economically-active individual in a type-h household. In other words, this is like the one-worker-per-household assumption which is common in much modelling, but the picture allows for unemployment.

Essentially, we have to introduce a set of weights which enable us to combine the utilities derived from the different aspects of life for which we have defined indicators above, and to subtract costs and relate these to income. Let I_{ij}^{m} be a composite indicator for the type-m individual living in i and working in j (and take $j = 0$ as representing 'unemployment'). We need to add further subscripts to specify house type l, job sector g and the set of services used, say s. Consider the following:

$$I_{ij}^{mlg} = Y_j^{mg} + \lambda^{(1)ml} a^l - p^{(1)l} + \sum_s \left(\lambda^{(2)ms} \sum_{j'} \frac{S_{ij'}^{ms}}{P_i^m} - p^{(2)s} \right) + \sum_p \left(\lambda^{(3)mp} \sum_{j'} \frac{T_{ij'}^{m*p}}{P_i^m} - m_{ij'} \right) \quad (5.43)$$

where a^l is a measure of the benefit to be derived from a type-l house and $p^{(1)l}$ is the unit cost of this, $p^{(2)s}$ is the cost of service s and m_{ij} is the money cost of transport. The terms $\lambda^{(1)}$–$\lambda^{(3)}$ are weights to help us combine the different utilities. The terms $S_{ij'}^{ms}/P_i^m$ and $T_{ij'}^{m*p}/P_i^m$ are measures of an individual's take-up of services s and trips of purpose p respectively. It would probably be simpler to measure this last component in terms of accessibilities X_i^{mp} with corresponding average costs $\bar{m}_i^p$.

The expression in (5.43) is a sum of weighted utilities plus residual income. It is interesting to try to simplify it to obtain a composite residential zone index for all type-m people. It will then be necessary to use the probability introduced earlier of whether someone is employed or not: p_i^{mg}. Then we can sum over job types and house types for m-residents of the zone:

$$I_i^m = \sum_{gj} p_i^{mg} T_{ij}^{mg} Y_j^{mg} + \sum_{gj} (1 - p_i^{mg}) Y_0^{mg} + \sum_l \left(\lambda^{(1)ml} a^l - p^{(1)l} \right) H_i^l + \sum_s \left(\lambda^{(2)ms} - p^{(2)s} \right) \sum_j S_{ij}^{ms} + \sum_p \left(\lambda^{(3)mp} - \bar{m}_i^p \right) X_i^{mp} \quad (5.44)$$

If it was possible to construct this index, then a per capita version

$$I_i^m / P_i^m \tag{5.45}$$

could obviously be computed. This can be taken as an index of 'residential zonal well-being'. This needs to be combined with an organizational index which will be tackled shortly. However, while the attempt presented above offers useful insights into the nature of the problem, it is clear that there are considerable research issues still to be pursued.

5.4 ORGANIZATION-BASED INDICATORS

We need to be concerned with efficiency and, in the case of marketed goods and services, profitability. But since we are concerned with spatial planning, we also need to find ways of representing how the scale of facilities in a particular place plays a role in the provision of goods or services to other sectors or to residents. Similarly, we need to investigate the role of organizations in the labour market. Thus, in summary, we have the main items of Table 5.1,

- efficiency of production
- role in the pattern of provision
- role in the labour market

and we consider each of these in turn.

5.4.1 Efficiency of production

Ideally, whether we are discussing organizations producing marketed or public goods or services, efficiency should be related to a detailed knowledge of the processes of production – in economic terms, to the production function. This, of course, is complicated by the effect of location and interactions with both suppliers and markets; and also because large organizations are themselves distributed in a variety of units across space. This detailed knowledge is rarely available and there is also a problem of scale. Ideally, as the title of the section implies, we need to focus on individual organizations – but effective data at this scale are rarely available. We have to make the best of what is available for 'sectors', probably zonal totals of some kind. We need to find indicators, then, which act as proxies for the production function: in a non-marketed sector volume of service supplied (e.g. hospital beds) per head of catchment population, or per head of resource inputs; in a marketed sector, the list extends to include indicators involving revenue, such as profit, profit per unit of input and so on.

Let W_j^g be the scale of production of g at j; let D_j^g be the revenue attracted (or some measure of usage for a public sector organization) and let C_j^g be the cost of providing the facilities. It should also be possible to calculate a catchment population (or a market segment), say Π_j^g. Then, useful indicators are

$$\frac{D_j^g - C_j^g}{C_j^g} \tag{5.46}$$

which is a measure of profitability, or simply

$$D_j^g / C_j^g \tag{5.47}$$

An efficiency indicator is the amount produced per unit cost:

$$W_j^g / C_j^g \tag{5.48}$$

and either production or costs can be examined in relation to catchment population:

$$W_j^g / \Pi_j^g \tag{5.49}$$

or

$$C_j^g / \Pi_j^g \tag{5.50}$$

5.4.2 Role in the pattern of provision

The problem to be tackled here is not one which has been seriously investigated in the literature (outside the consensus of the traditional narrowly defined equity–efficiency problem).

As we saw in Section 5.2 it is possible for organizations to be efficient without delivering effectively throughout a region. So it is necessary to calculate effective provision by delivery areas along the lines argued earlier:

$$\hat{W}_i^g = \sum_j \frac{T_{ij}^g}{T_{*j}^g} W_j^g \tag{5.51}$$

and we can then examine indicators such as

$$\hat{W}_i^g / P_i \tag{5.52}$$

In other words, we have to examine the variations in $\{W_i^g / P_i\}$ as the spatial pattern $\{W_j^g\}$ varies.

5.4.3 Role in the labour market

A similar argument can be carried through for job provision – indeed, we can think of organizations as producing 'jobs' as well as other goods and services, say E_j^g. $\{T_{ij}\}$ is the main labour market array introduced in Section 5.3.3. We can calculate

$$\hat{E}_i^g = \sum_j \frac{T_{ij}^g}{T_{*j}^g} E_j^g \tag{5.53}$$

and then an indicator such as

$$\hat{E}_i^g / \varepsilon_i P_i \tag{5.54}$$

(where ε_i is the economic activity rate in i) would be a measure of the proportion of desired jobs for residents of i which are offered by sector g. Again, we can examine how this indicator varies with $\{E_j^g\}$, and also how the probabilities of employment and unemployment vary.

5.5 SUITES OF INDICATORS FOR PUBLIC SERVICES

In the 'quality of life' literature it is public goods and services which have had the longest and widest exposure. This has blossomed in recent years in terms of a performance indicator perspective. As we saw in Chapter 2 this is partly a reflection of the Government (particularly UK) strategy of the mid-1980s which emphasized the perceived importance of measuring efficiency of performance in public services and nationalized industries. We can illustrate the use of indicators for public goods and services through a couple of examples.

5.5.1 Health care

Indicators for health care systems certainly pre-date the Department of Health and Social Security initiatives on performance indicators in the UK in the 1980s. Goldacre and Griffin (1983) provide an extensive review of indicators (and key statistics) in health care planning and remind us that as early as 1732 Dr Clifton, writing in *The Lancet*, was suggesting that certain items of data about all hospital inpatients should be systematically recorded, analysed and published to enable the work undertaken by hospitals to be assessed. Out of a plethora of references reviewed by Goldacre and Griffin, the work of Logan *et al.* (1972) provides 'one of the most comprehensive ranges of statistical indicators in the published literature'. The paper compares health services in Liverpool with those elsewhere, concentrating on a wealth of resource measures and activity measures. Table 5.2 lists the traditional sorts of indicator used in measuring activity rates, for example. This example illustrates a further geographical problem to which we will return: many of the indicators calculated involve ratios which must be modified to take account of cross-boundary flows.

In January 1982 the Secretary of State for Social Services in the UK announced arrangements for department reviews of each Regional Health Authority, and the first list of indicators was published later in 1982 (see Goldacre and Griffin 1983; DHSS 1984). The first package of performance

Table 5.2 Selected measures of activity and provision in health services

(i)	Admission rates per speciality per 1,000 population
(ii)	Average length of stays by speciality
(iii)	Throughput as measured by discharges per bed per speciality
(iv)	Percentage bed occupancy by speciality (occupied beds expressed as a percentage of available beds)
(v)	Operation rates per 10,000 population by speciality
(vi)	Admission to surgical beds per 10,000 population
(vii)	Ratio of admissions resulting in operations to all admissions in the surgical specialities
(viii)	Number of laboratory technicians and radiographers per 10,000 population
(ix)	Inpatient radiography units per 1,000 discharges
(x)	Provision of operating theatres per million population
(xi)	The number of discharges in surgical specialities per operating theatre
(xii)	The number of surgical beds per operating theatre

Source: Logan *et al.* 1972

Table 5.3 Indicators of clinical activity in the National Health Service

1	Urgent, immediate or emergency inpatient admissions per 1,000
2	All inpatient admissions per 1,000 population served
3	Average length of stay
4	'Throughput': average number of patients per head per year
5	'Turnover interval': average length of time a bed lies empty between admissions
6	Day cases as a percentage of deaths and discharges
7	New patients referred per 1,000 population served
8	Ratio of returning outpatients to new outpatients
9	Admission waiting lists per 1,000 population served
10	Estimated number of days taken to clear waiting lists at present level of activity
11	Percentage of obstetric admissions that result in births (including stillbirths)

Source: Pollitt 1985

indicators came under the major headings of clinical activity, finance, manpower, support service and estate management. Table 5.3 lists a new set of indicators concerned with clinical activity. Reviews of these sorts of indicators can be found in Pollitt (1985), Roberts (1990) and Mullen (1989, 1990).

In terms of a model-based indicator framework research in health care issues began in the mid-1980s (Clarke and Wilson 1984, 1985b). The importance of both residence- and facility-based indicators in formulating a new set of indicators based on patient flows was recognized. On the residence side we can first explore indicators relating to the proportion of residents treated within their own localities or receiving treatment some distance from home. These clearly relate to the spatial pattern of treatment of residents in a given area. From these it is an obvious next step to calculate the accessibility measures we saw in Chapter 3:

$$\frac{T_{ii}^k}{\Sigma_j T_{ij}^k} \tag{5.55}$$

(proportion of cases treated within a zone) and

$$\frac{T_{ij}^k}{\Sigma_j T_{ij}^k} \tag{5.56}$$

(proportion of cases treated in a surrounding zone) where T_{ij}^k is the flow of patients from residences in i to hospitals in j for specialty k.

A second set of indicators relate to hospitalization rates. These can be defined as the number of cases generated in a residential zone (by specialty) divided by the total population within that zone:

$$H_i^k = \frac{\Sigma_{ij} T_{ij}^k}{P_i^{(k)}} \ (\times 1{,}000) \tag{5.57}$$

where H_i is the hospitalization rate; P_i is the population in zone i or $P_i^{(k)}$ is some morbidity measure for specialty k.

Variations in such rates can be a function of a number of causes. Low rates may indicate limited availability of services, low incidence of that specialty, low referral patterns or high lengths of stay (thus reducing 'throughput'). Clearly, the indicators need careful investigation, and comparison between localities becomes a useful diagnostic measure.

Another set of residence indicators concerns the notional number of beds available per head of population (in days) and the notional expenditure on specialty k per head of population. The former can be expressed as

$$\text{nb}_i = \left(\sum_j \frac{T_{ij} a_j}{365} \right) P_i \tag{5.58}$$

where nb_i is the notional number of beds available to residents in i and a_j is the average length of stay (in days). The idea of notional number of beds results from the practical operation of the referral process between and within localities. It thus takes into account any cross-boundary flows. Because the indicator reflects actual referral patterns it becomes a meaningful indicator of the real accessibility of patients to facilities within and outside their home locality (and it may also partly explain the levels of hospitalization rates). Harvey *et al.* (1985) explore the relationship between notional bed availability and length of stay in terms of efficiency and effectiveness, arguing that a low

Table 5.4 The relationship between notional beds available to zone i and average length of stay

	High a_i	*Low* a_i
High nb_i	Inefficient Effective	Efficient Effective
Low nb_i	Inefficient Ineffective	Efficient Ineffective

Source: Harvey *et al.* 1985
Note: a_i is the average length of stay which patients from a residential zone i experience.

notional number of beds available to a residential area coupled with a short length of stay might indicate an efficient hospital but that the effectiveness of the service overall might be poor. These relationships are shown in Table 5.4.

Finally, we can add the effective provision indicator:

$$\hat{W}_i^k = \sum_j \frac{T_{ij}^k}{T_{*j}^k} W_j^k \tag{5.59}$$

where W_j might refer to beds available for specialty k.

On the facility side the concept of catchment population is again important. The crucial argument remains that the calculation of catchment populations is dependent on the level of provision in adjacent districts in addition to those available close by. As before, the catchment population can be calculated as

$$\Pi_j = \sum_i \frac{T_{ij}}{T_{i*}} P_i \tag{5.60}$$

Once this has been worked out then we can build up a suite of related indicators: for example,

$$B_j = \frac{W_i}{\Pi_j} \tag{5.61}$$

is beds per head of the catchment population;

$$E_j = \frac{C_j}{\Pi_j} (\times 1{,}000) \tag{5.62}$$

is budget (or expenditure) per head (or 1,000) of the catchment population; C_j is the expenditure in zone j cases per head of the catchment population.

Despite the progress made in performance indicators for health service planning it is fair to say that geography has largely been ignored by health service planners until very recently. The Working for Patients Act of 1989

(UK) has explicitly brought geography to the fore by splitting the purchasing of health care from the provision of health care. Health authorities, which traditionally looked after both roles, are now charged with purchasing health care facilities on behalf of their residents, theoretically at any hospital in the UK that can provide their services at the right price and at the least inconvenience. In turn, hospitals now have to compete with each other for patients and thus have to price and market their services accordingly. The result of this Act is likely to be a much more open and competitive market for health care in which both purchasers and providers need a better understanding of the geographical dimensions of health care needs, provision, utilization and outcomes. The implication of these changes for models and performance indicators is important. First, there is much more urgent need to examine the current performance of the market for health care (in terms of understanding patient flows by specialty) and therefore greater emphasis on catchment areas and effective provision indicators. Second, exploring scenarios concerning future purchaser–provider contracts allows the service planner to assess the likely outcomes through changes in the suites of indicators outlined above. On the residence side, indicators relating to accessibility and effective provision become much more significant as patient flows become less tied to nearby hospitals, whilst providers will increasingly be interested in market-based indicators of market size, penetrations and costs related to alternative scenarios based on changing service provisions (see Section 5.3 and Chapter 7).

5.5.2 Education

Work on indicators of education systems actually began in the late 1960s (Cohen 1967; Coombs 1969; Ferris 1969). Ferris, for example, was convinced that serious planning of the education system should always be accompanied by the development or use of indicators measuring the degree to which the goal of a plan is achieved (1969, p. 6). Little and Mabey (1972) examined the field of education and constructed an index for designating education priority areas using such indicators as teacher turnover rates, ability of children to speak English adequately, proportion of retarded, disturbed or handicapped pupils, poor attendance rates, truancy rates etc., all criteria originally recommended by the Plowden Report (1967) in relation to the need for public intervention in favour of areas where educational handicaps were reinforced by social handicaps (see also Johnstone 1978, 1981). Again, in terms of model-based research there has already been substantial progress in providing a framework for performance indicator calculations (Irwin and Wilson 1985).

The majority of these indicators were based on the facility side especially in relation to use of facilities and efficiency. They thus defined indicators based on pupil–teacher ratios, average class size, costs per subject taught, examination passes in different subjects and the size and ‘type’ of school catchment area. Given the latter indicator, it was then possible to define exam success per head of catchment population or indeed per head of residential population (effectiveness indicators). As with health care, recent legislation in education has put a new emphasis on school performance indicators. The Education Act of 1988 has brought choice and competition firmly onto the agenda with the abolition of fixed catchment areas and the publication of school performance indicators of exam success. These lists will supposedly inform parents of the success of different schools and allow them to make a free choice for their children (as long as places are available at the receiving school). Thus schools become directly competitive and must remain 'attractive' to parents in order to maintain rolls. If rolls decline then it is likely that future funding will also decline pro rata. The analogy with supermarket retailing has already been made in education systems outside the UK which have faced similar reform. As the impacts of the 1988 Education Act (UK) come into force, so a new range of indicators is required. These will need to relate both to school outcome measures (such as exam performance) and to the ability of schools to attract pupils. The former is important because current performance criteria are not adjusted to take account of pupil backgrounds or the nature of the residential environment in which they live (see the discussion in Douglas Willms (1992) and in Jowett and Rothwell (1988), for example, where the arguments are made without any explicit reference to the importance of geographical factors). In order to make progress let us define T_{ij}^{aw} as the number of pupils resident in zone i travelling to school in j (a, age group; w, social economic group); π_j as the catchment population for a school at j; and E_j^{st} as the number of 'passes' in subject s in year t.

Traditional indicators of exam success (and hence school effectiveness measures) are as follows: E_j^{st} itself; E_j^{st}/π_j (exam pass per head of catchment population);

$$\left(\sum_j \frac{T_{ij}^{**}}{T_{*j}^{**}} E_j^{st}\right) \Big/ T_{i*}^{**} \tag{5.63}$$

(pass rate per head of residential population). Given the importance attached to performance measurement (in terms of possible future funding) many would argue that at the very least we need to retain the disaggregation by 'pupil type', w, which would allow the indicators to be compared across pupil backgrounds and stop the unfair direct comparisons between inner city/ suburban school (much of the interest then shifts to comparisons between schools with similar intakes of pupil types). It might also be necessary to go one step further and attempt to include the effects of residential catchment neighbourhoods themselves (cf. Moulden and Bradford 1984; Bradford 1991): i.e. define

$$[T_{ij}^{awr}]$$

where r refers to residential neighbourhood characteristics. (The ongoing research is to define how that might be measured.)

Apart from fairer indicators of school effectiveness there is also a need to look at the effects of the reforms in terms of the likely impacts on the geography of school provision across a city as a whole. Parental choice (one of the driving forces behind the changes) is often not as 'free' as the rhetoric of the legislation suggests. There is a set of income and mobility constraints which tie people to localities (Bradford 1989). Those with greatest choice are likely to be middle/upper income earners. Therefore, schools with the best (currently unadjusted) performance scores might be in a strong position (especially if places are limited) to 'cream off' the more able pupils from the better backgrounds, leaving a residual pupil base in inner city areas or areas of lower income residents who effectively have no real choice over where they send their children. Thus indicators which keep abreast of changing catchment areas (by pupil or social type) will be crucial in monitoring the impacts of the 1988 Act. In addition, it will be interesting to model the growth/decline of schools over time as their 'attractiveness' waxes and wanes. If we define D_j as the number of pupils attracted to a school and K_j as the costs of supplying W_j places then the arguments of Harris and Wilson (1978) become relevant:

$$D_j > K_j W_j \quad \text{then } W_j \text{ will expand}$$
$$D_j < K_j W_j \quad \text{then } W_j \text{ will contract}$$

At equilibrium $D_j = K_j W_j$ for all j. The dynamics of indicators such as D_j/W_j and D_j/K_j then become particularly important to monitor.

The modelling and related policy tasks here are to work out how to make W_j more attractive for those schools who are faced with the downward spiral of falling rolls. Thus modelling methods may become as useful to individual schools as they do to monitoring city-wide changes. Given the discussion above, it is clear that schools, just like hospitals, will need to take a hard look at their existing service provision in the light of market shares and what the competition has to offer. Schools which find it hard to compete given exam performances *per se* may have to compete by offering more specialist tuition in the arts, sciences or even sports (given the confines of the National Curriculum). This again invokes images of retailers trying to maintain markets shares by 'product diversification' or niche marketing. If this is the case then school administrators will look increasingly to the performance indicators described in relation to the private sector (see Chapter 7).

5.6 AREA PROFILES

While the most important foci for performance indicators remain people and organizations, it is useful to develop the idea of area profiles. This enables us to compare the overall argument presented here with more traditional arguments, but also provides a useful backcloth for planning purposes: in many instances, decisions relate to *location* of facilities – and hence allocation to areas. We consider three headings:

- type
- role
- balance

and we examine each briefly in turn.

5.6.1 Type

Traditional analyses, as we have seen, have been focused on an areas 'type' in some way; and the apogee of this style of work was probably achieved in factor analysis. Such analyses will remain useful, especially immediately after a census with a mass of new data or to make a quick assessment of the structure of a city. In the argument developed here, we would expect to build up a 'type profile' simply by enumerating for each zone the main model variables: P_i^m, E_j^{gb}, Z_j^g and so on. (A useful addition would be measures of environmental quality.) It may also be possible to summarize 'type' by focusing on the range of household utility functions exhibited in each zone, though this to some extent overlaps with the idea of 'role' which we pursue next.

5.6.2 Role

We can further develop the concept of 'role' introduced above in relation to organizations. In effect, we summarized in relation to 'type' the people and organizations which make up different zones in a city. The next step is to use the interaction variables to assess the role a zone plays in the labour market, the service market and so on – thus focusing both on demand *and* production in each zone (see Chapter 8).

5.6.3 Balance

It is potentially useful to focus on the main inputs and outputs in each zone, particularly when the flows can be converted to economic (or at least common) units. In these terms, unemployment can be seen as the balance between labour supply and 'notional job' supply when viewed from the perspective of a residential zone; 'labour availability' can be taken as the difference between jobs and 'notional labour' supply when the area is viewed as an employment zone. In the case of money flows, it is likely that inputs and outputs – say wage income and consumer expenditure – should roughly balance. If such flows do not balance, then the difference may turn out to be a useful diagnostic indicator. This is a potentially valuable analysis, but it does leave substantial research questions in relation to the present state of the art.

5.7 INDICATORS AS DIAGNOSTICS

When all the indicators described above are computed, they provide a detailed portrait of people, organizations and areas. However, in some instances there may almost be too much information! The traditional way of dealing with this problem was to carry out something like a factor analysis which would, in effect, characterize areas in terms of groups of correlated variables. This was usually done in relation to data directly. The advantage of model-based performance indicators is that 'gaps' in data can be filled by model predictions.

The kind of analysis we are advocating here can be thought of as generating *diagnostic information*. It has already been demonstrated in the field of health services analysis (Clarke and Wilson 1985b) that there may be alternative ways of doing this which are useful. For example, take a set of indicators and for each calculate the mean and standard deviation across the analysis zones. We can then define an indicator as significant within a zone if it is $\pm x$ standard deviations away from the mean. The factor x can be fixed by the analyst: if it is large, very few zones will then be picked out, and vice versa. If these analyses can be executed as part of an on-line computing system, then it is possible to experiment with x values (which may in principle differ for each indicator) to generate good zonal diagnostics.

There is a second way in which new information can be gained from the basic set of indicators. This arises when it is possible to specify planning policies as targets to be achieved – say, for instance, in terms of 'notional shopping floorspace' per head of residential population (again see Chapter 2). For each zone for each such indicator, we can calculate 'percentage of target achieved'. It is then possible to search for zones whose indicators are outside the band 'target $\pm y\%$ of target', where, again, y is set by the analyst, ideally on the basis of on-line experiments. Experience with health care studies (Clarke and Wilson 1985b) shows this to be a powerful and informative procedure.

5.8 PERFORMANCE INDICATORS: COMPUTATION

In the previous section we articulated the sets of performance indicators which would provide a rich, valuable and diagnostic portrait of people, organizations and zones within a study area. It will already be clear that much of the data to compute what is ideally required is not routinely available. Much of the necessary information can be generated as model outputs: models can thus be used to fill gaps in data for the purpose of calculating performance indicators. However, one consequence of adopting a performance indicator perspective is a demand for foci and information which are not routinely available from models either. This raises research questions for the future of model building and design. To proceed, we first define the main variables more formally and then review the arrays which can be calculated.

5.8.1 The main variables

The most crucial decisions to take relate to levels of resolution. We need to categorize people and households, housing, the economy, marketed goods and services, public goods and services, and the elements of transport cost. At this stage, we have to balance what we would like to have against data availability and the ability of models to fill in any gaps. We carry through the argument with a minimum list.

- Individuals:
  - age
  - sex
  - terminal age of education
  - skills (occupation)
  - income
  - wealth
  - car ownership

together with specification of activities which will link individuals with the other sectors:

  - residential location
  - housing type
  - job type
  - demand for and pattern of use of services
  - transport costs
- Households
  - size and composition (number of adults, number of children)
  - aggregate income
  - aggregate wealth
- Housing
  - size
  - type
  - facilities available
- The economy
  - employment, by type (skills needed, wages and salaries etc.)
  - 'production'
  - scale of facilities – by sector
    - broad sectors (i.e. manufacturing, services etc.)
    - fine sectors (textiles, banking, retailing etc.)

(This includes a detailed specification of both marketed and non-marketed services.)

- Transport
  - networks
  - public provision – elements of generalized cost
  - trip purposes
- Area profiles
  - zoning system
  - environmental quality
- Connectivities
  - labour market
    - assignment of individuals to jobs
    - *wage rates*
  - housing market
    - assignment of households to housing
    - *prices*
  - service usage
    - assignment of individuals to units of
    - *service provision prices*

We have already developed an informal algebraic notation which we can summarize in relation to these lists as follows (bold italic type indicates a list of characteristics or vectors as appropriate).

- Individuals
  - characteristics $\boldsymbol{m}$
  - totals $P_i^{\boldsymbol{m}}$
- Households
  - characteristics $\boldsymbol{h}$
  - totals F_i^h
- Housing
  - characteristics $\boldsymbol{l}$
  - totals $H_i^{\boldsymbol{l}}$
- The economy
  - sectors g
  - employment characteristics $\boldsymbol{b}$
  - totals $E_j^{\boldsymbol{b}g}$
  - facilities characteristics $\boldsymbol{X}_j^g$
  - attractiveness W_j^g
  - production $\boldsymbol{Z}_j^g$
  - revenues D_j^g
  - costs of production $\boldsymbol{c}_j^g$
- Transport
  - modes k
  - characteristics $\boldsymbol{c}_{ij}^k$
- Area profiles
  - zones i, j
  - environmental quality $\boldsymbol{Q}_i$

- Connectivities
  - labour market $T_{ij}^{\boldsymbol{mbg}}$
  - housing market $R_j^{\boldsymbol{hl}}$
  - service usage $S_{ij}^{\boldsymbol{mg}}$
  - transport flows $T_{ij}^{\boldsymbol{mkp}}$

Needless to say, data are not available directly in relation to all these quantities, and certainly not at a fine level of resolution. Models can be seen as providing a capability to fill in gaps. The level of resolution will determine the techniques to be used. If the lists of characteristics $\boldsymbol{m}$ and $\boldsymbol{b}$, for instance, are of any length, then it will be necessary to use micro-simulation methods (see Birkin and Clarke 1988).

But it should also be recognized that it is not necessary to work at the same level of resolution throughout. It is more likely to be the case that several will be used simultaneously. It may be appropriate to use a coarse-scale comprehensive model to provide a backcloth and to cope with the main interdependences, a micro-simulation model for other purposes, and a series of single-sector studies at a fine scale to study, for example, the provision of particular services, health, education, retailing etc. It is against this background that we outline progress in Chapters 8 and 9 through a variety of what we might label first-stage applications.

6 Performance indicators for evaluation with a dynamic urban model

R. Tadei and H.C.W.L. Williams

6.1 INTRODUCTION

The aim of the preceding chapter was to develop a systematic and comprehensive framework for the conceptual identification and calculation of suites of performance indicators for use in a variety of situations in urban planning. The argument was constructed through the definition of efficiency and effectiveness indicators for a wide variety of subsystem models. These models and indicators were presented in a static framework although much of the interest in a planning context stems from monitoring changes in these indicators over time. Figure 6.1 illustrates how these indicators may be set in a comparative statics framework.

Although an urban system can be broken down into subsystems, the study of its dynamics implies not only the consideration of the dynamics of each subsystem taken separately but also the dynamics of the interrelationships between these subsystems. The aim of this chapter is to derive a set of economic indicators to be associated with a specific model of urban dynamics. The model is derived from economic principles which allow us to identify and relate a consistent set of indicators to the actions of individual consumers and producers in the markets for housing, labour and services. Indeed, a distinctive feature of the model is the derivation of the transition rates (or mobility choice) from random utility theory. A prerequisite to the establishment of economic measures within a dynamic framework will be a discussion of the formation of economic surplus measures in discrete choice contexts, and particularly those models of the competitive markets which are underpinned by probabilistic choice theory.

In Section 6.2 we remind the reader of the main components of the urban system. This is followed in Section 6.3 with an examination of the structure of the dynamic model to be used. In Section 6.4 we present the basic accounting framework used to discuss the importance of random utility models for the calculation of evaluation indicators described in Sections 6.5–6.8. Concluding comments are offered in Section 6.9.

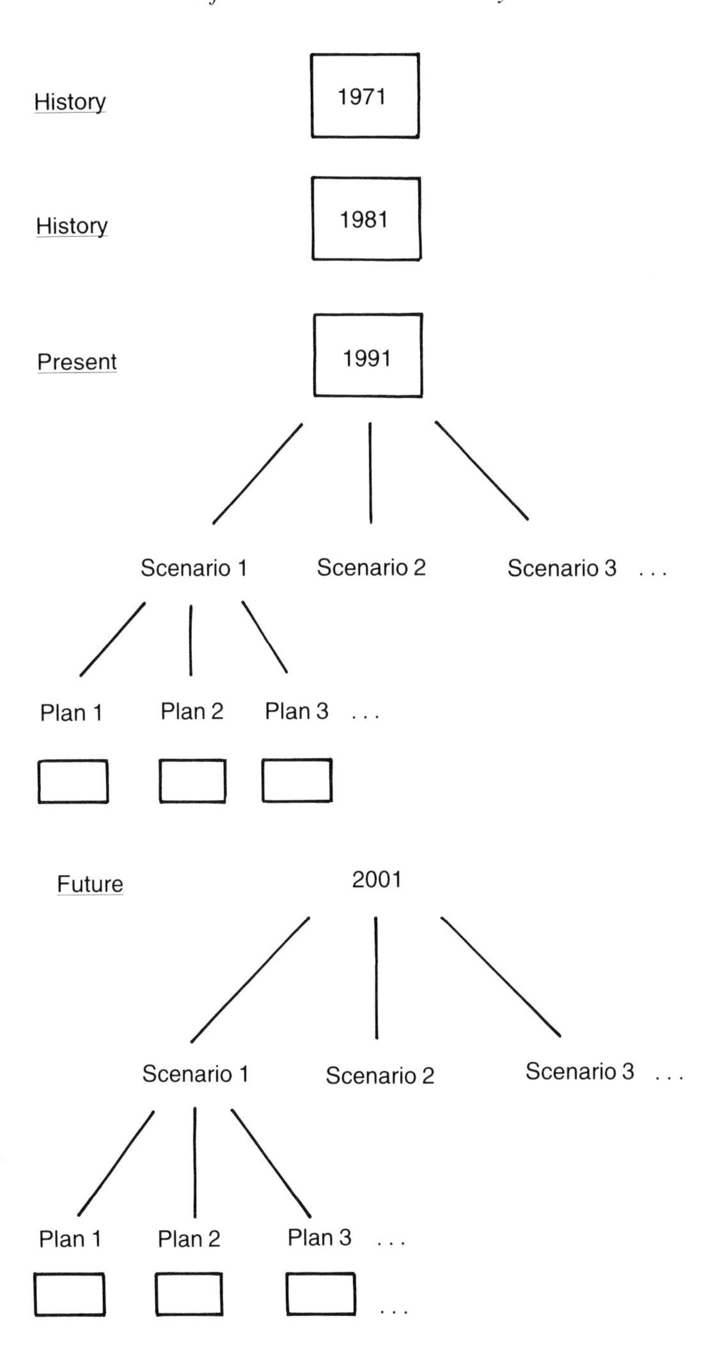

Figure 6.1 A structured set of entries

6.2 DEFINITION OF AN URBAN SYSTEM

An urban system can be seen as a set of elements (subsystems) interacting with each other through socio-economic and spatial mechanisms (Bertuglia *et al.* 1987b). The main subsystems of the urban system are

1. housing market
2. job market
3. service sector
4. land market
5. transport

and the corresponding variables for describing the structure of an urban system are as follows:

(a) population
(b) housing stock
(c) industry (economic base)

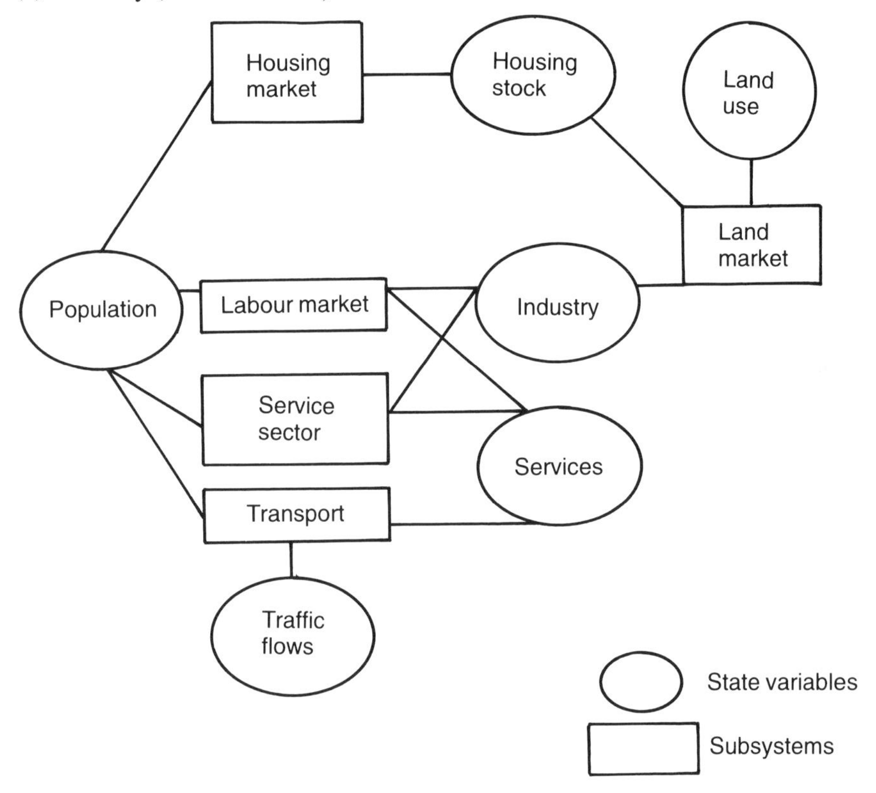

Figure 6.2 Main subsystems and states variables of an urban system and their interrelations
Source: Bertuglia *et al.* 1987b

(d) services
(e) land use
(f) traffic flows

The main interactions between these variables are shown in Figure 6.2.
The main subsystems shown in Figure 6.2 can be defined as follows.

1. The housing market subsystem is the set of relations between the population (housing demand) and housing stock (housing supply) creating phenomena such as residential mobility and house price dynamics.
2. The labour market subsystem is the set of relations between population (labour supply) and industries and services (labour demand) generating phenomena such as job mobility and wage dynamics.
3. The service sector subsystem includes the set of relations between population (service demand) and public and private service facilities (service supply) producing phenomena such as service location dynamics.
4. The land market subsystem is the set of relations between the building stock for residential, industrial and service use (land demand) and land supply giving rise to phenomena such as land price dynamics.
5. The transport subsystem is the set of relations between population and industrial and service activities producing phenomena such as journeys to work, journeys to services, commodity flows, modal demand and network flows.

6.3 THE STRUCTURE OF A DYNAMIC URBAN MODEL

Interest in dynamic models of urban systems has been influenced in recent years by a number of factors. First has been the reconstruction of processes of change in cities which has transformed their structure. These include the rapid decline of inner city populations, the growth of suburbs and small/medium sized towns, the relatively rapid loss of manufacturing jobs and the restructuring of employment around service occupations. All of these emphasize the importance of a dynamic framework.

Alongside these real structural changes have been attempts to incorporate dynamics within the modelling of urban and regional systems, with particular emphasis being placed on supply-side structures and the way these patterns evolve and change. This began with the work of Harris and Wilson (1978) on dynamics of retail change and was quickly extended into agriculture (Wilson and Birkin 1987), industry (Birkin and Wilson 1986a, b), education (Wilson and Crouchley 1984) and the joint consideration of retailing, residential location and housing (Clarke and Wilson 1985a). The work on applied retailing continued in Clarke (1986). Other researchers also made significant progress with a range of retail applications (Allen and Sanglier 1979, 1981;

Lombardo and Rabino 1983). The models, and the non-linear relationships within them, combined to generate bifurcations (fundamental structural change at critical parameter values), offering new insights into the nature of urban and regional change.

Another important contribution to the understanding of dynamics in urban systems was made by Weidlich and Haag (1983). Their approach is largely based on the theory of stochastic processes and master equations. Put simply, we can define $p(i, t)$ as the probability of finding the system in state i at time t. The evaluation of any probability distribution over time is determined by the conditional probability $p(i_2, t_2|i_1, t_1)$ defined as the probability of finding the system in state i_2 at time t_2 given that it was in state i_1 at time t_1. The master equation is simply a difference equation in time for such a conditional probability (Haag 1990).

The work on various systems (and on alternative model formulations) came together in two works on integrated urban models (Bertuglia *et al.* 1987b, 1990b). This integrated dynamic model for the city is based on three of the crucial relationships shown in Figure 6.2: the housing market; the market for services; the labour market.

The emphasis is on medium-term dynamics, focusing on processes which undergo changes over long periods (typically years) such as housing mobility, labour mobility and locational changes in services. The model combines the theory on dynamic systems from master equations with discrete choice models (Domencich and McFadden 1975; Anas 1982; Leonardi 1983, 1984) and bid–price markets (Ellickson 1981; Lerman and Kern 1983; Leonardi 1985). In addition, classic urban economic theory has been maintained for consistency (Alonso 1964; Beckmann 1973). In each time period the state variables measuring the quantities in the system (population, jobs, housing stock) are updated through transition probabilities. These are non-homogeneous (time dependent) and depend both on state variables and economic signals. Economic signals are computed according to the assumption that the associated prices clear the market (to ensure demand equals supply) within each period. On the demand side the aim is to maximize net utility whilst supply-side dynamics are driven by the desire of firms to maximize prices or revenues. The interaction mechanism between demand and supply is assumed to be a stochastic extremal process (Leonardi 1990a). Full details of the model are found in the Appendix.

The following sections describe a modified accounting framework for such a dynamic model of housing and labour and then the derivation of suitable evaluation (benefit) indicators associated with model outputs.

6.4 THE ACCOUNTING FRAMEWORK

Let us define the following variables: P is the total population in an area; Q_i is the total housing stock in zone i; S_j is the maximum number of possible workplaces in zone j; $T_{ij}(t)$ is the number of people living in i and working in j at time t; $E_j(t)$ is the number of new workplaces which can be opened in zone j at time t, with

$$E_j(t) = S_j - \sum_{i=1}^{n} T_{ij}(t)$$

$T_{i0}(t)$ is the number of unemployed people living in i at time t. $q_{ij}^{k}(t, \Delta)$ is the probability that a person living in zone i and working in zone j at time t will move to a dwelling in zone k at time $t + \Delta$; hereafter the assumption will be made that all transition rates refer to the interval $[t, t + \Delta]$. $q_{i0}^{k}(t, \Delta)$ is the probability that an unemployed person living in zone i will move to a dwelling in zone k; $p_{ij}^{k}(t, \Delta)$ is the probability that a workplace in zone j occupied by a worker living in zone i at time t will be occupied by a worker living in k at time $t + \Delta$; $p_{ij}^{0}(t, \Delta)$ is the probability that a workplace in zone j occupied by a worker living in i is closed down; $\rho_{ij}(t, \Delta)$ is the probability that a new workplace is opened in zone j and occupied by a worker living in zone i.

The accounting equations are now presented for both the employed and the unemployed, and take the standard form:

(number in a state at time $t+\Delta$) = (number in the state at time t) + (number moving into the state) – (number moving out of the state) + (components arising from additions or subtractions from the stock of jobs or residences)

In terms of the transition rate definitions, the accounts for the employed become:

$T_{ij}(t+\Delta) = T_{ij}(t) \times (1 -$ residential moves from i to any k – loss of job at j and replacement by unemployed living anywhere
\+ residential moves into i for workers currently employed in j + all currently unemployed in i who find jobs in j at the expense of those living in any k
\+ the component arising from new workplaces being created)

This is given by

$$T_{ij}(t+\Delta) = T_{ij}(t)\left[1 - \sum_{k=1}^{n} q_{ij}^{k}(t,\Delta) - \sum_{k=0}^{n} p_{ij}^{k}(t,\Delta)\right]$$
$$+ \sum_{k=1}^{n} T_{kj}(t)\left[q_{kj}^{i}(t,\Delta) + p_{kj}^{i}(t,\Delta) + E_j(t)\,\rho_{ij}(t,\Delta)\right] \quad (6.1)$$

The accounts for the unemployed involve the following transitional components: residential moves for the unemployed, both into zone i, and out of zone i;

workers living in i who become unemployed; those unemployed in i who become employed; a job creation component. The accounts thus become:

$$T_{i0}(t + \Delta) = T_{i0}(t)\left\{1 - \sum_{k=1}^{n} q_{i0}^{k}\right\} + \sum_{k=1}^{n} T_{k0}(t)q_{k0}^{i}(t, \Delta) + \sum_{j} T_{ij}(t) \sum_{k=0}^{n} p_{ij}^{k}(t, \Delta) - \sum_{k=1}^{n}\sum_{j=1}^{n} T_{kj}(t)p_{kj}^{i}(t, \Delta) - \sum_{j=1}^{n} E_{j}(t)\rho_{ij}(t, \Delta) \quad (6.2)$$

6.5 HOUSING AND LABOUR MARKET SPATIAL INTERACTION MODELS

6.5.1 Introduction

In this section we consider the basis for spatial interaction models within the framework of discrete choice theory, and more specifically focus on various housing and labour market models derived from random utility theory. Our main purpose is to contrast previous model development with our labour market model, with a view to reformulating the latter to bring about more realism and consistency with existing work.

The models presented in Bertuglia *et al.* (1990b) develop the doubly constrained spatial interaction model (DCSIM) (or generalizations of that form) within residential and employment location choice contexts. It has been known for more than a decade that the DCSIM may be derived as a model of residential choice with supply-side (housing stock) constraints; a model of job choice with (employment) constraints; or a joint choice of residence and job within a combined housing and labour market context.

The essential common features of these approaches are as follows:

1. identification of a set of discrete choices, and the decision-making unit(s) concerned (e.g. individual, household, firm etc.);
2. establishment of the basis for option selection and the essential criterion (preference) function defined over the choice set;
3. identification of the representative and random components of the preference function and specifying the distribution of the random variables;
4. invoking random utility calculus to derive probabilistic choice models;
5. developing subsidiary arguments for market interactions (e.g. establishing differential rent levels through market clearing).

In all these approaches an attempt is made to explain the dispersion of trip behaviour (and in the case of the journey to work, the locational behaviour) in terms of differences in 'observable' and 'unobservable' attributes of decision-making units. In the following cases we shall be concerned with homogeneous market segments from the point of view of the former, and attribute different locational choices to dispersion in the structure of preferences and to unobserved factors in the utility functions used to record preferences. Heterogeneity in the observable attributes (household structure classes, demographic and socio-economic characteristics) may be easily incorporated through the market segmentation process. The limitations of the 'pure choice' and competitive market approach are well known and will not concern us here. An identification of possible extensions is usually made with the model derivations.

Our purpose now will be a brief consideration of distinct existing forms from the point of view of the steps identified above, considering first the third approach – the joint choice derivation. We shall refer to all these derivations in the subsequent (re)formulation of the general dynamic model.

6.5.2 The joint choice of residence and job

Such an approach has been implicit in the literature since the early 1970s but received explicit and extensive treatment a decade later in two papers by Tanner (1980) and Anas (1983). We shall focus on the interpretation in Tanner's treatment and refer only briefly to the derivation by Anas. (Problems of zonal aggregation bias are ignored in the following.)

Tanner introduces rent and salary differentials between the zones. Thus, for a home in zone i, the rent is taken to be R_i, and in zone j a mean salary level of S_j is defined, in the same units. (The values of the vectors $\boldsymbol{R}$ and $\boldsymbol{S}$ will be manipulated until each zone becomes sufficiently attractive to fill the appropriate numbers of houses and jobs.)

In the choice of residence or workplace, an individual is assumed to take into account not only rents, salaries and travel costs, but also other aspects of her/his personal preferences in relation to houses and jobs. These are represented by random components in the costs/utilities associated with each house/job pair and are drawn from probability distributions.

The choice postulated is one in which each of the N individuals chooses jointly where to live and work, and does this by finding the particular house and job which offers her/him the largest net utility, and solves the following optimization problem:

$$\max_{\substack{\text{over alternatives} \\ (i,j)}} \quad (S_j - R_i - c_{ij} - \varepsilon_{ij}) \tag{6.3}$$

From the usual assumptions of identical and independently distributed (iid) Weibull distributions, the multinomial logit model follows immediately, giving the expected number of trips between zones i and j as

$$T_{ij} = \frac{NO_iD_j \exp[\lambda(S_j - R_i - c_{ij})]}{\Sigma_{ij}O_iD_j \exp[\lambda(S_j - R_i - c_{ij})]} \quad \text{for all } (i,j) \tag{6.4}$$

where c_{ij} is the travel cost from zone i to zone j and S_j and R_i take values to satisfy the trip end (market clearing) conditions

$$\sum_j T_{ij} = O_i \qquad i \in I$$
$$\sum_i T_{ij} = D_j \qquad j \in J \tag{6.5}$$

Equation (6.4) will be written in standard form as

$$T_{ij} = a_i O_i b_j D_j \exp(-\lambda c_{ij}) \tag{6.6}$$

with

$$a_i = \frac{\exp(-\lambda R_i)}{(D/N)^{1/2}} \tag{6.7}$$

and

$$b_j = \frac{\exp(\lambda S_j)}{(D/N)^{1/2}} \tag{6.8}$$

with

$$D = \sum_i \sum_j O_i D_j \exp[\lambda(S_j - R_i - c_{ij})] \tag{6.9}$$

There is a well-known indeterminacy of the values of the rents and wages which may be computed up to an arbitrary constant. As Tanner (1980) notes:

> the assumption that rents and salaries are transfer payments excludes the possibility that any element of them represents intrinsic desirability of particular locations rather than the monetary transactions reflecting accessibility to houses and jobs.

(This will emerge in the economic evaluation to be considered later.)

Williams (1976) has shown that the appropriate performance (evaluation) indicator for this model is given (in the above notation) by

$$\sum_i O_i \log a_i + \sum_j D_j \log b_j \tag{6.10}$$

Important points to note are that individuals choose residences and jobs through a process of maximizing their preferences, subject to their income constraints, and that the housing price and wage differentials are determined through market clearing at the residential and employment locations respectively.

6.5.3 Models of the labour market

In this section we provide a brief extension to the approach adopted by Anas (1987) to include transportation costs following the approach of Train and McFadden (1978), within the classical 'goods–leisure trade-off' framework. Later, we shall combine these features within the present framework.

In the labour market model proposed by Anas (1987), individuals choose the region and occupation in which they want to supply their labour and also how many hours they wish to supply their labour. They may also choose to remain unemployed and receive government benefits. Firms in each region (or zone) choose the amount of labour they need to hire from each occupation.

We first consider the 'continuous–discrete' labour market model, the former descriptor referring to the determination of the hours worked, the latter to the location–occupation. We define the following variables:

j	location–occupation labels referring to the choice set $[J, j = 1,\ldots, n]$
L	leisure hours
t	the number of hours worked
H	total hours available for work or leisure
C	other consumption expenditures
c_{ij}	cost of transportation to and from work
W_j	wage rate
h	unearned income (non-wage income)
$U(L,C)$	the utility function used to record preferences over different combinations of leisure and consumption.

The number of hours worked is determined from the simple process of maximizing utility, subject to budget and time constraints:

$$\max_{L,C} U(L,C) \tag{6.11}$$

subject to

$$\text{budget constraint} \qquad B(C, W, c) = 0 \tag{6.12}$$

$$\text{time constraints} \qquad T(H, t, L) = 0 \tag{6.13}$$

with non-negativity restrictions on the variables.

Substituting the values for C and L in the objective function our aim is to maximize U over the hours worked, yielding a quantity $V^*(H, c, W)$ which is dependent on the wage rate in each location–occupation combination. Train and McFadden have shown the form that this function takes under different structures for the utility function, and specifically for logarithmic and Cobb–Douglas forms (Anas adopts the former).

With such a utility function the individual can now make a choice between the discrete options. One of the options is unemployment which may be denoted by U, in which case $L = H$ and the household receives the government benefit B as compensation. Introducing random utility terms $\hat{e}$ and e which are uncorrelated with the choice of C and L, then under the usual assumptions regarding the distribution of random variables the multinomial logit model gives the probability that an individual will choose to work (or be unemployed) in region j. Anas goes on to derive the demand function for labour as a function of the wage rate W_j, from profit-maximizing assumptions. The total demand for labour in any occupation–region is then obtained (in terms of total hours). Market clearing in the labour market equilibrium model for the static case is then expressed through the derivation of 'excess demand functions', or, in short, by equating demand to supply in each region–occupation, thus providing a set of simultaneous equations for the wage rates given all other constants as exogenous.

In the model of Leonardi (1990b) (see the Appendix) a different approach is taken in the formulation of the discrete choice process. It is specifically stated:

> One should not forget that in the labour market the roles of households (or more precisely, working householders) and entrepreneurs is interchanged relative to the housing market. In terms of modelling style, this means that we attribute to households a utility maximizing behaviour while we attribute to entrepreneurs a revenue maximizing behaviour and a price (rent) setting option. Moreover, we insist on all demand units being satisfied (i.e. everybody must be somewhere) while some supply units might remain vacant (i.e. some dwellings may remain empty). In a labour market the working householders (both employed and unemployed) are the supply, while the firms are the demand. In terms of modelling style, this means that we attribute to householders an income maximizing behaviour (a close analogue of revenue maximizing) and a wage setting option (a close analogue of price setting) while we attribute to entrepreneurs a utility maximizing behaviour (utility in this case being often interpreted as productivity or profit). We still insist that all demand units be satisfied (which now means that all jobs must be filled), while some supply units might remain vacant (which now means that some householders might be unemployed).
>
> (pp. 329–30)

In contrast to the model of Anas, the model of Leonardi has the following features.

1 Both firms and households influence wages through an explicit bidding process, and the equivalent market clearing conditions at equilibrium are derived as a by-product rather than used as a starting assumption.
2 The total demand for labour is fixed, i.e. there is a given list of job positions available in each zone – this amounts to ignoring problems of production technology and production functions. Here, for illustration,

each job will be considered as given and performed by some worker, and the only type of decision that the firm can take is to choose from a list of applicants to the job the one yielding the highest utility to the firm.

3 The household is assumed to have no other sources of income besides its wages.

In this model the utility each job (i.e. firm) attaches to each worker is typically a measure of the profit each worker can produce, i.e. the difference between the revenue and the wage paid to him/her. It is assumed for simplicity that revenues have no systematic relationship with workers of different groups, i.e. workers are basically homogeneous in skill and productivity – on the average. However, heterogeneity in skill and productivity is allowed for in a stochastic fashion by adding a random term to the firm's utility function.

The formulation of the model then proceeds by writing down a typical utility for a job $h \in \Gamma_j$, the set of jobs in zone j, of a worker $z \in \Omega_i$, the set of households in zone i,

$$u_{zh}(t) = \varepsilon_{zh} - w_{zh}(t) \qquad z \in \Omega_i, \quad h \in \Gamma_j \tag{6.14}$$

in which ε_{zh} is the random utility term and $w_{zh}(t)$ is the market wage to be paid for a worker living in zone z and working in zone h at time t.

It is assumed that each household meets the usual budget constraint, such that his/her expenses for the consumption of transportation, housing and services must be less than his/her wage. Introducing a slack variable (disposable income) we have

$$w_{zh}(t) = c_{ij} + e_i + R_i + y_z(t) \tag{6.15}$$

with c_{ij} the commuting cost between zones i and j, e_i the average expenditure on services (including transportation cost) from zone i, R_i the average housing expenditure for a household living in i and, by definition, $y_z(t)$ the disposable income for a household z at time t.

It is then stated that the commuting costs, service expenditure and housing expenses can be considered as deterministic variables, common to all (h, z) for each (i, j) pair, while the disposable income is in general a random variable (because the wage is a random variable).

Defining v_{ij} as the expenses incurred by an individual living at i and working at j

$$v_{ij} = -(c_{ij} + e_i + R_i) \tag{6.16}$$

then the utility function is written as

$$u_{zh}(t) = v_{ij} + [\varepsilon_{zh} - y_z(t)] \qquad z \in \Omega_i, \quad h \in \Gamma_j \tag{6.17}$$

The discussion then proceeds (cf. Leonardi 1990b):

> Some comments of clarification on the budget equation are useful. The fact that the wage is expressed as a sum of different consumptions and a disposable income does not mean that the firm offering the job $h \in \Gamma_j$

> is aware of the detailed consumption list of the worker $z \in \Omega_i$. The only information exchanged in the bidding process between firm and worker is the whole wage w_{zh}. However, the worker knows the composition of his budget constraint, and he negotiates his wage with the firms in such a way as to get the highest possible value for y_z, the disposable income. The variable y_z should therefore be interpreted as the highest disposable income a worker z can get from the market. This is the result of a maximization the worker carried out over a whole list of jobs he/she has information about up to time t and doesn't depend on a specific job. This is why it is labelled by the subscript z only. There is a close analogy between the disposable income y and the dwelling price r in the housing market. Both are obtained through a process of bid maximization. . . . In the labour market a worker receives bids from the firms, i.e. the highest wage they are willing to pay for a job, and through the budget constraint, the corresponding disposable incomes are computed and the 'price' is set equal to the maximum among such incomes.
>
> (Leonardi 1990b, pp. 334–5)

What this implies is that in order to attract workers to j from i a wage $w(i, j)$ must be offered to compensate for the additional transport component to render the disposable income independent of job location. Each worker is subject to a budget constraint related to the place of residence, so that he would not be willing to accept a wage which does not cover his expenditure. Therefore the wage a firm in zone j pays for a worker living in zone i is

$$w_{ij}(t) = [c_{ij} + c_i(t) + R_j(t)] + y_i(t) \tag{6.18}$$

where $c(t)$ is the total cost of service trips and the utility function for a firm hiring a worker living in zone i (the profit) is

$$v_{ij}(t) = \Pi_j(t) - w_{ij}(t) \tag{6.19}$$

where $\Pi_j(t)$ is the firm's revenue for a job at j.

Examination of equation (6.18) clearly reveals that the hypothesis underpinning the choice process will dictate which of the real wage or the disposable income is a dictating variable. Clearly one or both will be dependent on i and j. If wages are determined at the employment zones as in the previous model then w is j-dependent and the disposable income is necessarily dependent on i and j. If on the other hand the bid wages by firms are realistically dependent on i and j then it is possible, as in the present model, to allow the disposable income to be dependent only on the residential label.

It is certainly not necessary to assume that, because the households and firms 'reverse roles' from the demand and supply viewpoints, the hypotheses must be such as to retain some sort of close analogy between the behaviour in the two markets. Although the process considered here is consistent it is not very realistic. It would seem appropriate to think of wages being determined by market clearing, as for example described by Anas (1987). This results in a j dependence for the wages and an i and j dependence for the disposable income (which may, or may not, be the criterion the individual job seekers are attempting to maximize).

It therefore seems appropriate to modify the labour market model in such a way as to marry some of the features of the conventional 'job choice' approach with some of the features of the present approach.

6.6 A REVISED LABOUR MARKET MODEL

The following features characterize the reformulation of the labour market model. Individuals are assumed to select a job from a discrete set ($j = 1,\ldots, n$) on the basis of maximum utility (or, alternatively, maximum disposable income); second, in any given period an individual will either be in work or be unemployed – there will be no consideration of the number of hours worked; third, in a given time period a fixed stock of jobs exists – no explicit consideration will be given to the production technology of the firms. Lastly wages will be determined through the usual market clearing assumptions which equate demand to supply.

We shall write the net (representative) utility of choosing a job in zone j (given a residence in zone i) as

$$v_{ij} = U_{jw} - c_{ij} - C_i - R_i \qquad (6.20)$$

in which we define U_{jw} as the maximum utility derived from a job in zone j offering (an average zonal) wage $w(j)$, c_{ij} as the cost of a return trip between zones i and j, C_i as the total cost of service trips from the residential zone i, R_i as the residential rent in zone i (which in the labour market model is given and not subject to endogenous determination) and W_j as the prior probability that an individual will select a job in zone j (discontinuing price variables).

Introducing a random term (iid Weibull) to account for dispersion of preferences, unmeasured attributes and in this case zonal aggregation effects, the probability p_{ij} that an individual will select j is given by

$$p_{ij} = \frac{W_j \exp(\lambda v_{ij})}{\Sigma_j W_j \exp(\lambda v_{ij})} \qquad (6.21)$$

An index $j = 0$ may refer to a state of unemployment with an associated (representative) utility of U_0, which includes the unemployment benefit. The quantities W_j represent the number of jobs in any zone, and thus W_j/W_* is the *a priori* probability that an individual will select a job in zone j in the absence of travel considerations. In the stochastic assignment models of Leonardi (1990a) the weight is given in terms of the number of opportunities which become available and are recorded by an individual.

The total number of trips between i and j (given one trip per worker) is then given by

$$T_{ij} = P_i p_{ij} \tag{6.22}$$

in which P_i is the number of workers residing in zone i.

In this model the market is cleared, and an equilibrium wage distribution over the zones is determined by equating the supply of labour to a zone, as a function of the relative prices, with the number of jobs:

$$\sum_i P_i p_{ij} \leq Q_j \tag{6.23}$$

In the Leonardi model the spatial interaction models are formed through a process of mutual bidding in which the market clearing is performed directly. To establish a link with that model we introduce the quantity ψ_j. It can be established that $\exp(\psi_j)$ represents the probability that a supply unit remains vacant and $1 - \exp(\psi_j)$ is the probability that it is assigned. The market clearing conditions then become

$$\sum_i P_i \frac{W_j \exp(\lambda v_{ij})}{\Sigma_j W_j \exp(\lambda v_{ij})} = Q_j[1 - \exp(\psi_j)] \tag{6.24}$$

We are now in a position to consider the dynamic extensions of the model and in particular the models and indicators for the transition rates.

6.7 MODELS AND INDICATORS FOR THE TRANSITION PROBABILITIES

6.7.1 Introduction

In Section 6.3 we considered the formation of random utility models of discrete choice in the labour and housing markets at a single point in time. The dynamic extensions are cast in a simple framework in which the choice set is remain in the present state or transformation to any one of the available known states. Logit models are derived in terms of the representative utilities associated with these states.

6.7.2 Transition probabilities and market clearing for the joint choice model

The probability of making a transition between the states (i, j) and (k, l) is given by

$$\text{prob}\,[(i,j) \rightarrow (k,l)] = \frac{W(k,l)\ \exp[\beta U(k,l)]}{\Sigma_l\ \Sigma_k\ [W(k,l)]\ \exp[\beta U(k,l)]} \tag{6.25}$$

in which $U(k, l)$ is the representative utility associated with the new state. In order to emphasize the current state the appropriate term may be split off from the denominator.

The transition to states which only involve a change in one label, and are therefore confined to either the housing or the labour market, follows immediately, and the probability of remaining in the present state is simply prob$[(i,j) \rightarrow (i,j)]$, or 1 minus the probability of transitions to all other states:

$$p(i,j \rightarrow i,j) = 1 - \sum_{\substack{k \quad l \\ (k,l) \neq (i,j)}} \sum p(i,j;k,l)$$

$$= \frac{\exp[\beta U(i,\ j)]}{\sum_{k} \sum_{l} \Delta W(k,l)\ \exp[\beta U(k,\ l)] + \exp[\beta U(i,\ j)]} \quad (k,l) \neq (i,j) \tag{6.26}$$

The expected maximum utility associated with this set of choices (to remain in the current state or make a transition to any other state) is

$$\langle \hat{U}_{ij} \rangle = \frac{1}{\beta} \log\{\exp[\beta U(i,j)] + \sum_{k} \sum_{l} \Delta W(k,l) \exp[\beta U(k,l)]\} \tag{6.27}$$

When the time increment is small this takes the form

$$U(i,j) + \frac{\Delta}{\beta} \sum_{k} \sum_{l} W(k,l) \exp\{\beta [U(k,l) - U(i,j)]\} + O(\Delta^2) \tag{6.28}$$

and in the limit as $[\Delta]$ tends to zero it attains the expected utility value for the state (i,j). When, in these models, we take the limiting form of $\langle \hat{U}_{i,j} \rangle$ as $\Delta \rightarrow 0$ and attain a value such as (i,j) corresponding to the 'base state', it is important to realize that $\langle \hat{U}_{i,j} \rangle$ relates to that subpopulation who actually select the option (i,j) in the first place. This, in principle, is different from the representative utility for the whole population *prior to that choice being made*. Corresponding to zero transitions from that state the increase in total expected utility due to residential and employment mobility decisions is thus expressed in terms of the second term as before.

6.7.3 Market clearing

It is a fundamental assumption of any of the incremental dynamic models considered that all markets are cleared within one period, i.e. the prices (rents and wages) at the end of each period, say at time $t + \Delta$, are computed in such a way as to balance demand and supply as determined by conditions set at the beginning of the period. Thus, the market clearing calculations require a determination of the demand and supply of houses and jobs, respectively, in the relevant time increment.

The total demand for housing in zone k is given by

$$H_k(t, \Delta) = \sum_i \sum_j P_{ij}(t) p(i, j \rightarrow k, j) \tag{6.29}$$

and the supply of houses in the present model by $W_k(t)$ or, in the Leonardi formulation, by

$$W_k(t) \{ 1 - \exp[-\phi_k(t)] \} \tag{6.30}$$

The total supply of *labour* to zone j is, similarly,

$$L_j = \sum_i \sum_k P_{ik} p(i, k \rightarrow i, j) \tag{6.31}$$

and the total supply of jobs will vary according to which model is used. In the simplest form, the supply of jobs will be given in terms of turnover plus new jobs (or minus jobs removed). In either model, the current version or Leonardi's (which includes the $1 - \exp(\psi_j)$ terms) two equations for demand and supply of housing and labour must be solved simultaneously. In this model there effectively exists a single 'market' for residences and jobs within which combinations of options are sought and prices are determined (at the relevant location) to equate supply and demand.

6.7.4 Transition models and market clearing for the revised model

In the present model the choices are made in separate markets and it is an assumption of the models that in the time interval $[t, t + 1]$ only transitions involving a change in job or residence can occur. We emphasize that, in the reformulated model, the original restriction that mobility in the labour market arises from unemployed workers taking the jobs of workers who have moved or left can be relaxed.

In the joint choice model, the quantity $p(i, j \rightarrow k, l)$ is the joint probability of making a residential transition from i to k and a job move from j to l. With due regard to the definition of choice sets, as defined in Section 6.4, the probabilistic choice model for the residential transitions conditional on employment in zone j becomes

$$q_{ij}^k = \frac{\lambda \Delta W_k(t) \exp[\beta U_{kj}(t)]}{\exp[\beta U_{ij}(t)] + \lambda \Delta \Sigma_k W_k(t) \exp[\beta U_{kj}(t)]} \tag{6.32}$$

in which $\lambda \Delta W_k$ is the number of vacant dwellings in zone k known to the demand unit in the time interval $[t, t + \Delta]$, U_{kj} is the representative utility associated with working in j and living in zone k, $k = 1, \ldots, n$, and β and λ are non-negative constants. The expected maximum utility indicator associated with this set of choices is thus

$$\tilde{U}^{R}_{*j} = \frac{1}{\beta} \log \left\{ \lambda\Delta \sum_{k \neq i} W_k(t) \exp[\beta U_{kj}(t)] + \exp[\beta U_{ij}(t)] \right\} \quad (6.33)$$

which for small Δ may be written

$$\tilde{U}^{R}_{*j} = U_{ij}(t) + \frac{\lambda\Delta}{\beta} \exp[-\beta U_{ij}(t)] \sum_{k \neq i} W_k(t) \exp[\beta U_{kj}(t)] \quad (6.34)$$

As Δ tends to zero this measure tends to the utility of the initial state (i, j). The increment above this value associated with the possibility of residential mobility is given by

$$\delta\tilde{U}^{R}_{*j} = \frac{\lambda\Delta}{\beta} \sum_{k \neq i} W_k(t) \exp\{\beta[U_{kj}(t) - U_{ij}(t)]\} \quad (6.35)$$

which clearly vanishes as the time interval tends to zero.

Corresponding relations may be derived for the labour market. In terms of the quantities V_{ij} (the utility function for a firm at j which employs a worker from zone i), P_i (the total number of unemployed persons who live in zone i) and μ, β (positive constants) the probability that a firm at j will release an employed worker living in i in preference to an unemployed worker living in k is given by

$$p^{k}_{ij} = \frac{\mu\Delta P_k(t) \exp[\beta V_{kj}(t)]}{\mu\Delta \Sigma_k P_{k0}(t) \exp\{[\beta V_{kj}(t)] + 1\} + \exp[\beta V_{ij}(t)]} \quad (6.36)$$

The expected maximum utility indicator associated with this set of choices is given by

$$\tilde{U}^{\mathrm{E}}_{ij} = \frac{1}{\beta} \log \left(\mu\Delta \left\{ \sum_{k \neq i} P_{k0}(t) \exp[\beta V_{kj}(t)] \right\} + 1 \right) + \exp(\beta V_{ij}) \quad (6.37)$$

and the net increase of utility in the interval above the initial state at time t becomes

$$\delta \tilde{U}_{i*}^{E} = \frac{\mu \Delta}{\beta} \left(\sum_{k \neq i} P_{k0}(t) \exp\{\beta[V_{kj}(t) - V_{ij}]\} + \exp(-\beta V_{ij}) \right) \quad (6.38)$$

6.8 EVALUATION INDICATORS FOR THE INTEGRATED MODEL

6.8.1 Introduction

Evaluation indicators of interest include

1 the economic surplus measures associated with the set of mobility choices of transitions and
2 the economic evaluation measures associated with occupancy of current states.

We note that significant complications occur with the present model owing to the interdependence of the mobility decisions associated with the common components of utility functions.

6.8.2 Conditional housing and labour market utility models

As emphasized in Bertuglia *et al.* (1987b), it is a condition of the mobility models for the housing and labour markets that the employment or housing state respectively remains constant. This conforms with the assumption that only one transition – either a residential move or a job change – may occur in the time interval. The formation of such conditional probabilities is given in Bertuglia *et al.* (1990b). In the original notation, we can write, with explicit forms for the three choice models, for residential mobility

$$\begin{aligned} q_{ij}^{k} &= \tilde{p}_{j}^{i} \tilde{q}_{ij}^{k} \\ &= \frac{\lambda \Delta W_{k}(t) \exp[\beta(U_{kj} + V_{ij})]}{D_{ij}} \end{aligned} \quad (6.39)$$

and for labour mobility

$$\begin{aligned} p_{ij}^{k} &= \tilde{q}_{i}^{j} \tilde{p}_{ij}^{k} \\ &= \frac{\mu \Delta P_{k}(t) \exp[\beta(U_{ij} + V_{kj})]}{D_{ij}} \end{aligned} \quad (6.40)$$

The denominator D_{ij} is given by

$$D_{ij} = \left[\exp(\beta U_{ij}) + \lambda \Delta \sum_{k'} W_{k'}(t) \exp(\beta U_{k'j}) \right] \left\{ \exp(\beta V_{ij}) + \mu \Delta \left[\sum_{k'} p_{k'}(t) \exp(\beta V_{k'j}) + 1 \right] \right\} \tag{6.41}$$

To first order in Δ the expressions may be written

$$q_{ij}^{k} = \frac{\lambda \Delta W_k(t) \exp[\beta(U_{kj} + V_{ij})]}{D_{ij}} \tag{6.42}$$

$$p_{ij}^{k} = \frac{\mu \Delta P_{k0}(t) \exp[\beta(U_{ij} + V_{kj})]}{D_{ij}} \tag{6.43}$$

in which the denominator D_{ij} takes the form

$$D_{ij} = \Bigg(\exp[\beta(U_{ij} + V_{ij})] + \Delta \Bigg\{ \lambda \exp(\beta V_{ij}) \sum_{k'} W_{k'} \exp(\beta U_{k'j}) + \mu \exp(\beta V_{ij}) \left[\sum_{k'} P_{k'}(t) \exp(\beta V_{k'j}) \right] \Bigg\} \Bigg) \tag{6.44}$$

It seems reasonable to expect that the limiting forms of the evaluation measure will tend to the utility of the base states as the increment of time, Δ, tends to zero.

The formation of the evaluation indicators involves summing all the expected utilities associated with all choices made, including remaining in a given state or making a transition to any other available state. If the workplace and residential decisions were truly independent then the total evaluation measure could be written in terms of benefit components associated with these states: V_{ij}, in which V_{ij} is the mean value for those who choose to remain working in state ij, and U_{ij}, in which U_{ij} is the mean value for those who choose to remain resident in zone i.

$$\text{Utility associated with those who remain in the same state} = U_{ij} + V_{ij} \tag{6.45}$$

For the evaluation procedure it would be interesting to know whether we can find a function, or functions, Y such that partial derivatives with respect to appropriate utility components will generate the transition matrices and thereby the demand for housing and employment.

Given the prominence of the 'log denominator' in the consumer surplus expressions for the multinomial logit models, does this expression play a similar role for those conditional mobility models which are not exactly of logit form? Given the expression

$$D_{ij} = \exp[\beta(U_{ij} + V_{ij})] + \lambda\Delta \exp(\beta V_{ij}) \sum_{k'} W_{k'} \exp(\beta U_{kj})$$
$$+ \mu\Delta \exp(\beta U_{ij}) \left[\sum_{k'} P_{k'} \exp(\beta V_{k'j}) + 1 \right] + O(\Delta^2) \quad (6.46)$$

it may readily be shown that the following expression holds for the residential mobility transition rates:

$$q_{ij}^{k} = \frac{1}{\beta} \frac{\partial}{\partial U_{kj}} \log D_{ij} \qquad k \neq i \quad (6.47)$$

For labour mobility the corresponding expression is

$$p_{ij}^{k} = \frac{1}{\beta} \frac{\partial}{\partial V_{kj}} \log D_{ij} \quad (6.48)$$

The probability of remaining in the same residential and labour market state is

$$\text{prob(no moves)} = \tilde{p}_j^i \tilde{q}_i^j$$
$$= \frac{\exp[\beta(U_{ij} + V_{ij})]}{D_{ij}}$$
$$= \frac{1}{\beta} \frac{\partial}{\partial U_{ij}} \log D_{ij} + O(\Delta) \quad (6.49)$$

The expression $\log D_{ij}$ thus appears to have an important role in evaluation.

We now consider the total zonal demand for housing and employment and define H_k^j as the demand for housing in zone k by those currently working in zone j and L_k^i as the demand for employment (labour) in zone k by those currently living in zone i. Then

$$H_k^j(t, \Delta) = \sum_i P_{ij}(t)\, q_{ij}^k \quad (6.50)$$

$$L_k^i(t, \Delta) = \sum_j P_{ij}(t)\, p_{ij}^k \quad (6.51)$$

Now in terms of

$$Y_{.j} = \frac{1}{\beta} \sum_i P_{ij} \log D_{ij} \quad (6.52)$$

$$Y_{i.} = \frac{1}{\beta} \sum_j P_{ij} \log D_{ij} \quad (6.53)$$

then

$$\frac{\partial Y_{.j}}{\partial u_{kj}} = \sum_i P_{ij} q^k_{ij}$$

$$= H^j_k \tag{6.54}$$

and

$$\frac{\partial Y_{i.}}{\partial v_{ik}} = \sum_j P_{ij} p^k_{ij}$$

$$= L^i_k \tag{6.55}$$

Thus $Y_{.j}$ and $Y_{i.}$ are the evaluation functions which generate the appropriate zonal demands *due to transitions* when differentiated by the appropriate utility components.

The quantity

$$Y_{..} = \frac{1}{\beta} \sum_{ij} P_{ij}(t) \log D_{ij}(t, \Delta) \tag{6.56}$$

is thus suggested as the *total evaluation index* for transitions in the housing and labour markets.

This single function thus appears to contain sufficient information to express the total associated with the residential and employment mobility decisions. But, does Y have an appropriate limiting form as $\Delta \rightarrow 0$? It is readily shown that this limit is given by

$$Y_{..}(\Delta \rightarrow 0) = \sum_{ij} P_{ij}(t)\,(U_{ij} + V_{ij}) \tag{6.57}$$

If the utility components U_{ij} and V_{ij} were independent (appropriate to independent housing and labour market decisions) and contained no common elements, then the utility of the base state would be $U_{ij} + V_{ij}$ and $Y_{..}$ would be an acceptable limiting form.

However, the derivation of economic evaluation measures for the constrained model considered here is more complicated than the joint choice model considered in Section 6.4 because the models for residential choice and job choice are not independent – although the choice contexts are specified separately. Thus, the same total utility function U_{ij} characterizes the relative preference for the state (i, j) and this is merely 'broken up' when considering the individual transitions – i.e. for residential moves all workplace-specific components of the utility function will be factored out, and conversely for the job choice component. There is thus the danger of double counting the common utility components if we see these two decisions (of where to live given workplace and where to work given residence) as independent.

This does suggest, however, that we then should explore additions of the measure Y which generate the appropriate limiting form. Defining the quantities

$$\begin{aligned} U_{ij} &= U'_{ij} - c_{ij} \\ V_{ij} &= V'_{ij} - c_{ij} \end{aligned} \tag{6.58}$$

then an appropriate limiting form would be

$$\dot{Y}_{..}(\Delta \rightarrow 0) = \sum_{ij} P_{ij}(t)\{U'_{ij} + V'_{ij} - c_{ij}\} \tag{6.59}$$

$$= \sum_{ij} P_{ij}(t)\{U_{ij} + V_{ij} + c_{ij}\} \tag{6.60}$$

$$= Y_{..}(\Delta \rightarrow 0) + \text{total transport cost} \tag{6.61}$$

This argument suggests that the appropriate evaluation form should be

$$Y_{..} = \frac{1}{\beta} \sum_{ij} P_{ij}(t) \log D_{ij}(t, \Delta) + \text{total transport cost} \tag{6.62}$$

It should be noted that if prices or wages are included directly in the exponents (the v and u functions) then (to bring about market clearing) these should be treated as transfers and appropriate terms should be added to the evaluation function.

6.9 CONCLUSIONS

It is appreciated that the discussion above is complex, and this of course reflects the complexity of the dynamic urban models which underpin the arguments. The important conclusion is that it is possible to define appropriate measures to evaluate the range of alternative frameworks described above in relation to different assumptions concerning the behaviour of individuals and firms. This is clearly important when we are faced with many different hypotheses about the way in which housing and labour markets actually function. One research task is thus immediately apparent: the calculation of the evaluation indicators for a real city using the different assumptions highlighted above. This is a difficult task given the wide range of data requirements but it is one which would be ultimately rewarding. In addition, we have shown that despite the complexity of these models it is possible to redefine the outputs of the models to produce powerful indicators for measuring and comparing alternative assumptions concerning the transitions in housing and labour markets.

7 Understanding retail interaction patterns: the case of the missing performance indicators

M. Birkin

7.1 INTRODUCTION

In recent years we have witnessed a growth in the use of spatial interaction models in a private sector planning context (e.g. Wilson 1988; Clarke and Sinclair 1991; Birkin *et al.* forthcoming). The main objective of this chapter is to explore the implications of this switch in emphasis for performance analysis.

Consider, as a starting point, the performance indicators for marketed goods and services outlined by Clarke and Wilson in Chapter 5. Broadly, the suggested indicators are as follows:

Catchment population of a centre
Service availability
Benefit, or consumer surplus
Total revenue attracted
Residential travel costs
Service availability per head
Benefit per head

There are two points to be made about this framework. The first is that all the indicators are based around an interaction variable of the form

$$\left\{ S_{ij}^{mg} \right\}$$

where i and j are locations, g is the type of good or service and m is a person-type.

This does not really give us the information we need in a competitive situation, which will almost by definition involve an interest in variables of the form

$$\left\{ S_{ij}^{kmg} \right\}$$

where k is a competing provider of goods of type g. So, for instance, this now allows us to calculate a model-based revenue for providers at j as $\Sigma_{im} S_{ij}^{kmg}$.

Table 7.1 Performance indicators for private sector planning

Residence-based indicators	*Facility-based indicators*	*System-wide indicators*
Market penetration	Catchment population	Average distance travelled
Average distance travelled	Group performance	Market share
Provision ratio	Market share	...
	...	

(Readers unfamiliar with this style of analysis may be wondering about the usefulness of this type of indicator, as most businesses can be expected to have a good idea of the turnover of their own outlets – but this should become apparent later on.)

The second point which we notice about Clarke and Wilson's list is that most of the emphasis is on variation in the indicators by i and m for particular products g. In other words, they are primarily measures of equity: how does performance vary for different types of individual at a location; and for the same types of individuals at different locations.

For practical purposes, business users of a modelling approach will be far more interested in efficiency-based indicators of the j–k variety for particular products: how is the business performing at different locations *vis-à-vis* its competitors? So, our next step is to define a more appropriate set of indicators – see Table 7.1. Let us note immediately that we are ignoring cost- and profitability-based indicators for present purposes, because in our experience these are rarely introduced at the present level of the analysis. More usually these will be introduced at a later stage in the evaluation process. So what we are concerned with are explicitly geographical indicators.

We have retained a conventional distinction between residence-based, facility-based and system-wide indicators. The comments here should not be taken to imply that residential variations are of no interest. For example, market penetration is repeatedly seen to be a more meaningful measure of spatial variations in a company's hold in a particular sector than the centre-based market share, and examples are provided in Section 7.2 below.

However, this should not be seen as a definitive typology. It may also be useful to make a different type of distinction between

- indicators of business performance
- indicators of model performance
- summary type indicators

The first types of indicators will be used to add a spatial dimension to more traditional measures of business performance. For example, if we use a production-constrained spatial interaction model (see Chapter 3) to generate a predicted pattern of spatial interaction flows, then we can sum these flows across all residential areas to gain model-based estimates of the potential revenue of a facility. We can divide such predicted revenues into known facility turnover to give 'group performance', an index of how well a facility performs against expectations. Group performance of greater than unity indicates that a store is a good performer, while performance of less than unity indicates a poor performance. These indicators will often reward further examination.

Another useful indicator of business performance is market penetration: the proportion of expenditure in a given residential area which is accounted for by competing institutions. It is well known in the UK grocery sector, for example, that Sainsburys and Gateway have much higher market penetration in the south than in the north, while companies such as Kwik Save and Morrisons are much stronger in the north. Typically, however, one can see variations which are equally great within a region; and again there may be an obvious pattern to this, with some retailers stronger in the towns, perhaps, and others strongest in suburban or rural areas. Figure 7.1 shows examples of this for the UK car market. The large and systematic variation in market share across regions of the UK is mirrored by variations in the market penetrations between postal districts within a single region.

Indicators of model performance are crucially important in validating the way in which the model reproduces system behaviour. There are two types of issue which deserve mention here. One is that obvious measures of model performance include things like the average travel distance or the regional market share of the different players – indicators which can be verified against market intelligence or survey data. The issue is that the indicators often used may be too crude for the job, and we return to this in Section 7.3. The second issue is that indicators like revenue attracted to a facility or group performance may function as indicators not only of business performance but also of model performance. So a major objective in modelling is to obtain a set of revenue predictions for individual stores which closely match known totals to provide confidence in the model; yet we also know there will be variations in performance from place to place which we must not lose sight of. This is an important and problematic tension in modelling.

Summary indicators are not direct measures of business or model performance but may be used to enhance the understanding of what is going on in a region; and thus perhaps to explain the patterns in other key indicators. For example, we can generate estimates of the catchment population of centres and also their revenues. If we look at sales per head of catchment population then we would expect a good correlation between this and branch performance. Towns with high sales per head of catchment population and no branches currently might suggest themselves as good potential locations in an expanded network (in common parlance, they are 'undershopped'). A simple indicator like centre size may be used as an additional filter in this kind of exercise (Table 7.2).

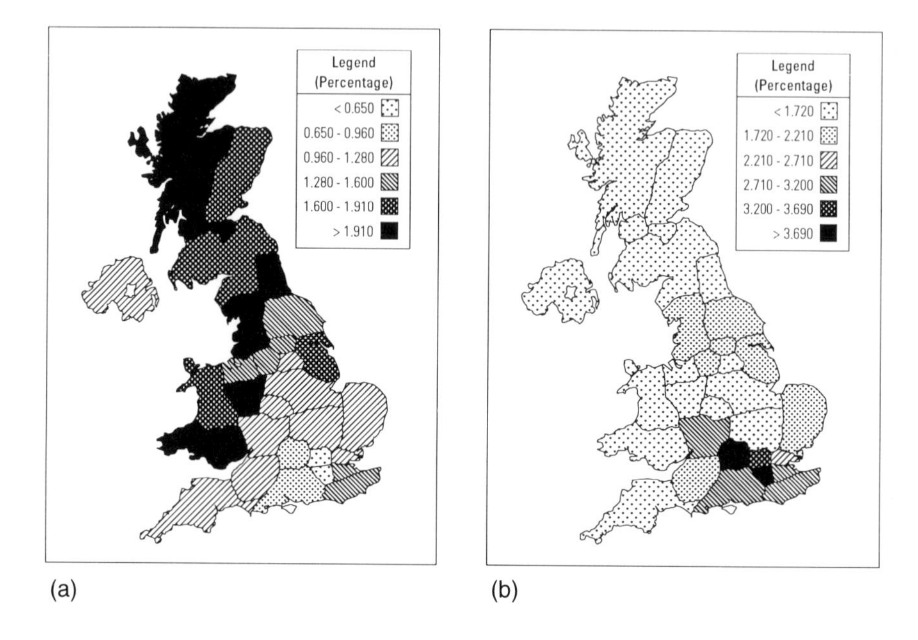

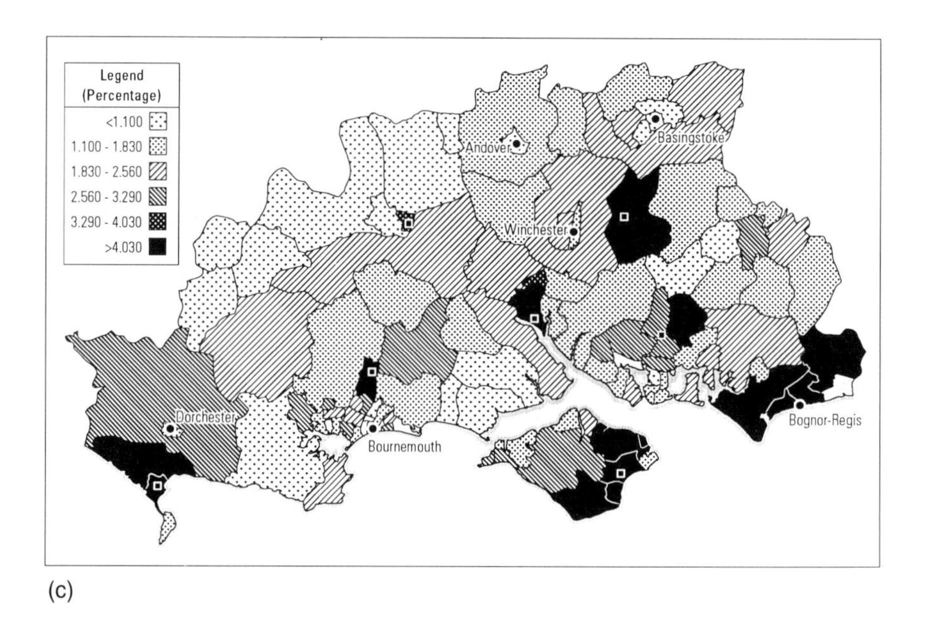

Figure 7.1 (a) UK regional market penetrations for Lada; (b) UK regional market penetrations for BMW; (c) market penetrations for a leading motor distributor by postal district, with dealer locations (■)

Table 7.2 Revised categorization of performance indicators

Business		*Model*	*Summary*
Market penetration		Average distance travelled	Centre revenue
Centre market share	←	Regional market share	Catchment population
Group performance	→		Sales per head of catchment population
			Centre size
			Catchment composition

7.2 A FRAMEWORK FOR NUMERICAL SIMULATION

The simulations in this chapter are based on a simple singly-constrained spatial interaction model which takes the form

$$S_{ij}^{kg} = A_i e_i P_i W_j^{\alpha} \exp(-\beta c_{ij}) \qquad (7.1)$$

and

$$A_i = \frac{1}{\Sigma_j W_j^{\alpha} \exp(-\beta c_{ij})} \qquad (7.2)$$

In this case, we take the i-zones to be the set of postal districts in West Yorkshire; j is a set of fifteen major centres within the region; k relates to three imaginary competitors for a single product ($g = 1$), white electrical goods. The expenditure estimates have been derived from survey data in such a way that average expenditure varies according to the age and social class mix of the postal districts. An imaginary pattern of floorspace provision has been established in which Leeds, Bradford and Wakefield are the largest centres. The competitors numbered 1 and 2 are both national multiple retailers which have large stores in the major centres only. Competitor number 3 is a local chain with smaller outlets in a larger number of centres. The model uses a β value of 0.2, yielding an average travel distance of around 5 miles.

Some performance indicators from the baseline model are shown in Figures 7.2 and 7.3 and Table 7.3. Figure 7.2(a) shows a relatively balanced pattern of market penetrations across the region for the national multiple which only falls away in very rural areas. This pattern is sustained through large and attractive stores in the major centres of the region. However, the greatest peaks are around Ilkley and Otley where there are outlying branches. Figure 7.2(b) shows a much more up-and-down pattern for the local chain. Performance is rather weak in urban areas, where the retailer is out-competed by national chains, but is stronger in suburban and rural areas. Figure 7.3 shows the expected pattern of average distance travelled by consumers across

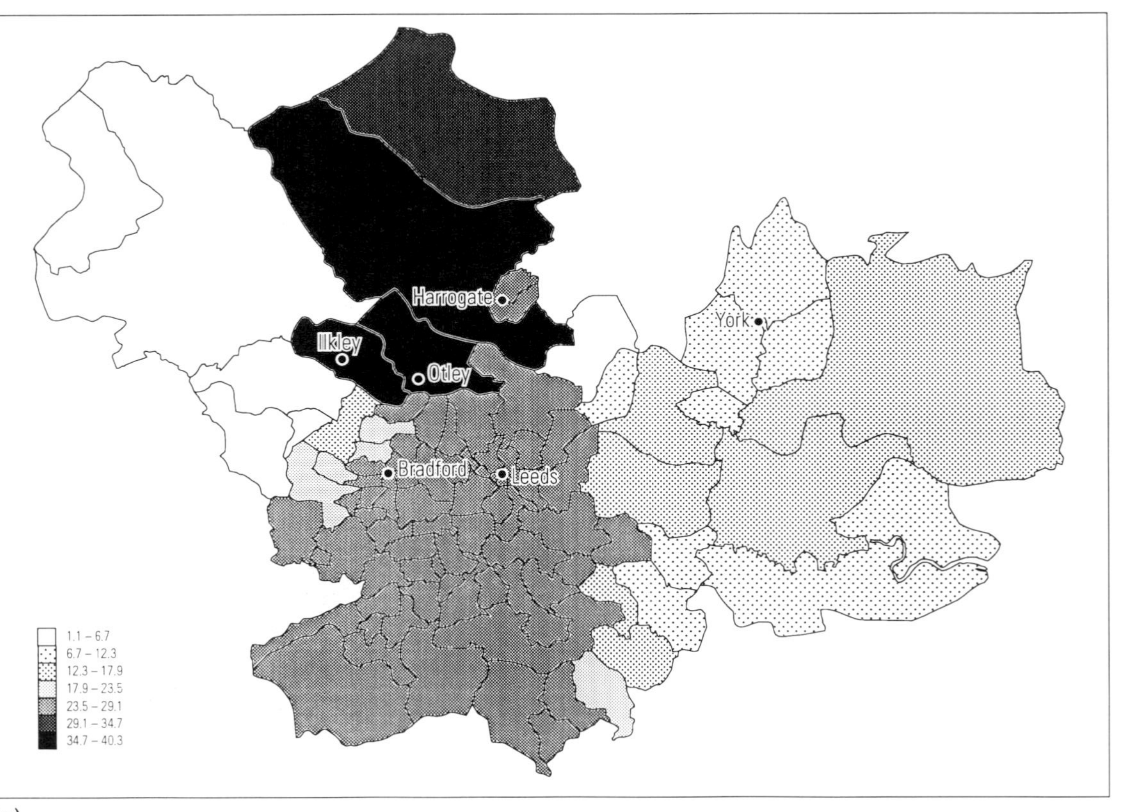
Harrogate
York
Ilkley
Otley
Bradford
Leeds
1.1 – 6.7
6.7 – 12.3
12.3 – 17.9
17.9 – 23.5
23.5 – 29.1
29.1 – 34.7
34.7 – 40.3

(a)

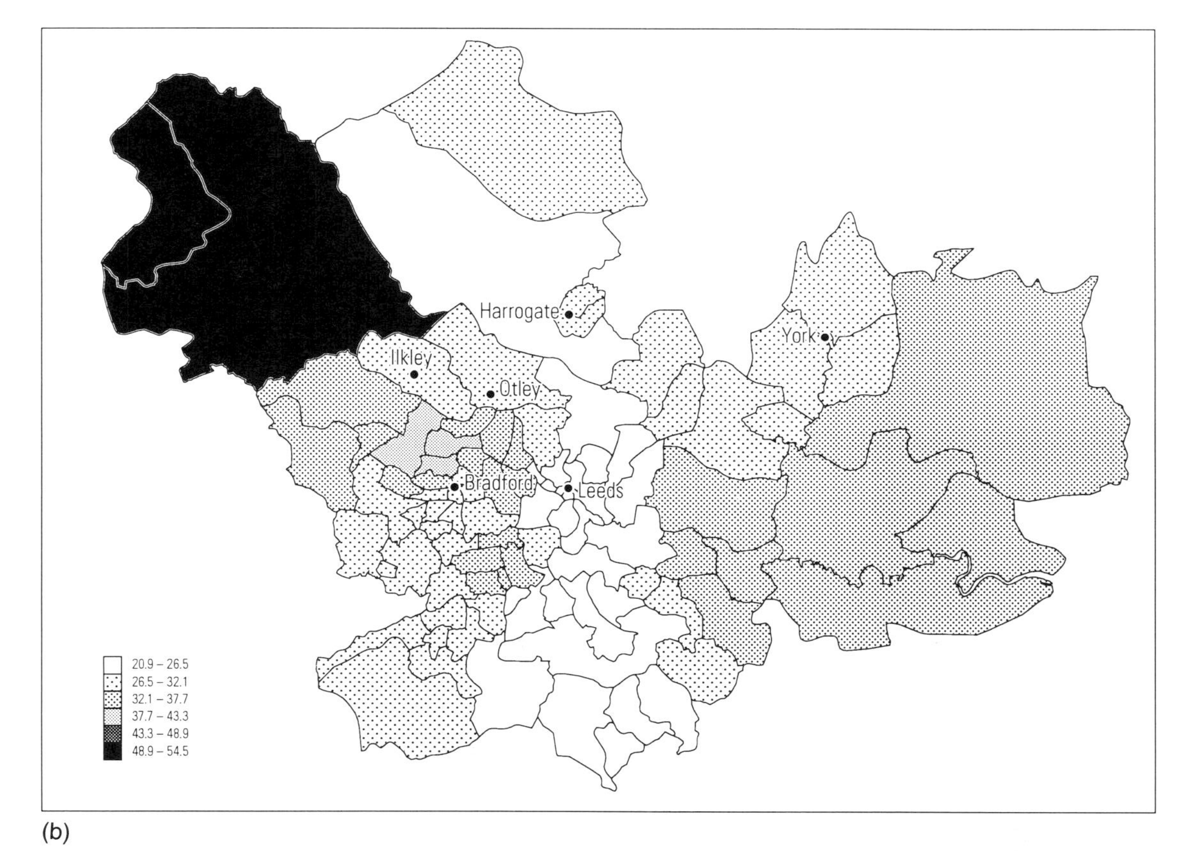

(b)

Figure 7.2 (a) Market penetrations for a national chain; (b) market penetrations for a local multiple

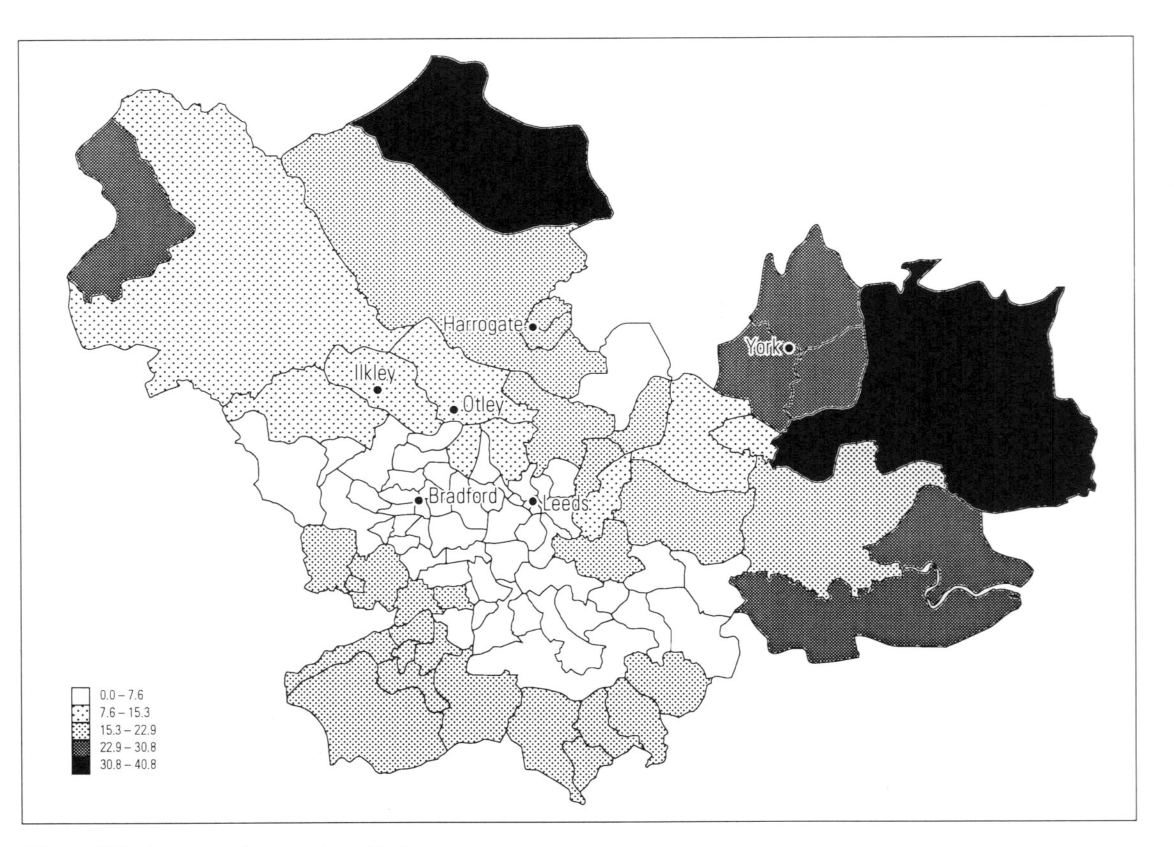

Figure 7.3 Average distance travelled

Table 7.3 Catchment populations of the major West Yorkshire centres, with an imaginary distribution of outlets

1	Batley	43,666
2	Dewsbury	197,978
3	Pontefract	94,491
4	Castleford	90,030
5	Wakefield	249,475
6	Wetherby	39,159
7	Ilkley	20,201
8	Otley	33,186
9	Horsforth	11,569
10	Pudsey	21,559
11	Leeds	564,289
12	Skipton	33,638
13	Shipley	38,311
14	Keighley	88,175
15	Bradford	557,089

the region. Table 7.3 shows the catchment populations of the centres in the region.

7.3 SOME NEW INDICATORS

In the previous section, we have argued that in moving from the analysis of public sector service planning into private sector applications, it is not generally sufficient to transfer performance indicators and modelling methods uncritically. In this section, we consider the weaknesses associated with particular indicators, and suggest alternatives which may be more appropriate.

7.3.1 Residence-based indicators

A key indicator here is effective delivery, generally defined as

$$\frac{\Sigma_j \; W_j S_{ij}}{D_j} \tag{7.3}$$

where D_j is the revenue accruing to centre j. This indicator aims to achieve a notional distribution of facilities proportionately between residents who actually use those facilities. This is particularly useful when the facility provided can be expressed as a divisible commodity such as hospital beds.

However, problems arise in retail applications. In this situation, the spatial interactions will usually be dependent, in a non-linear way, on the 'attractiveness' of retail centres, and conventionally

$$W_j = Z_j^{\alpha} \tag{7.4}$$

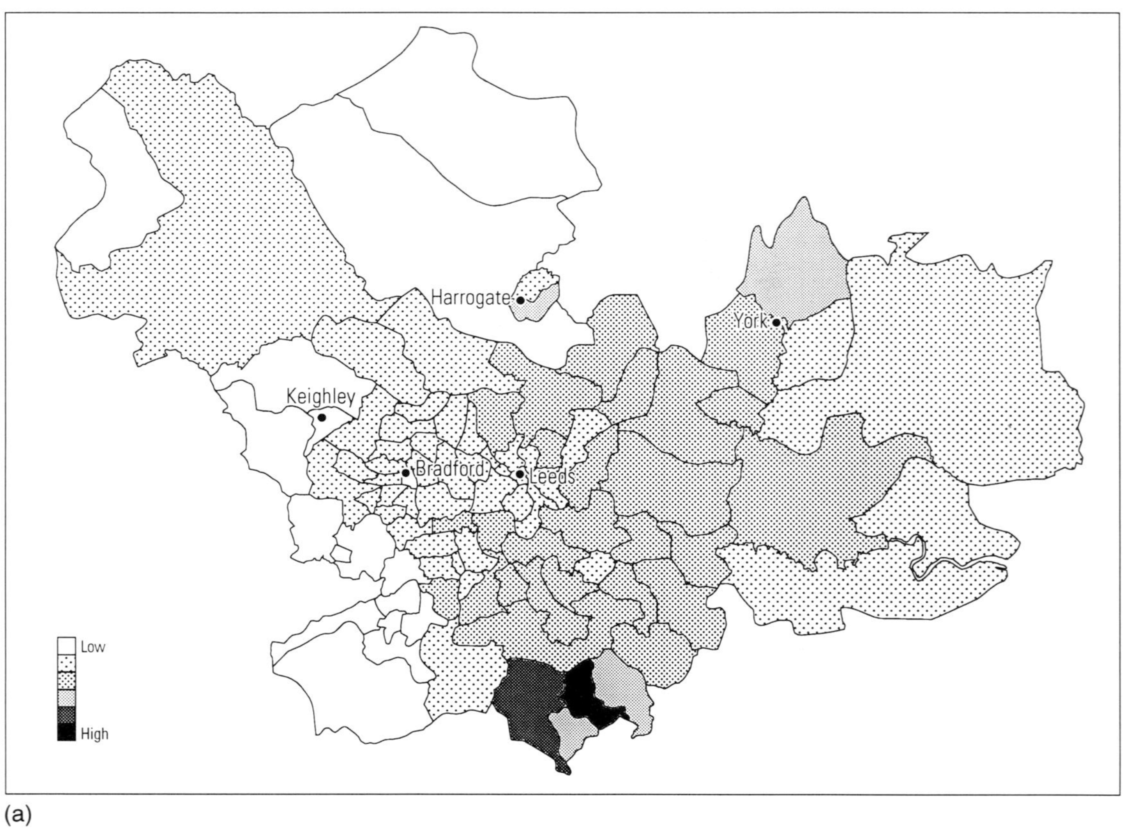

(a)

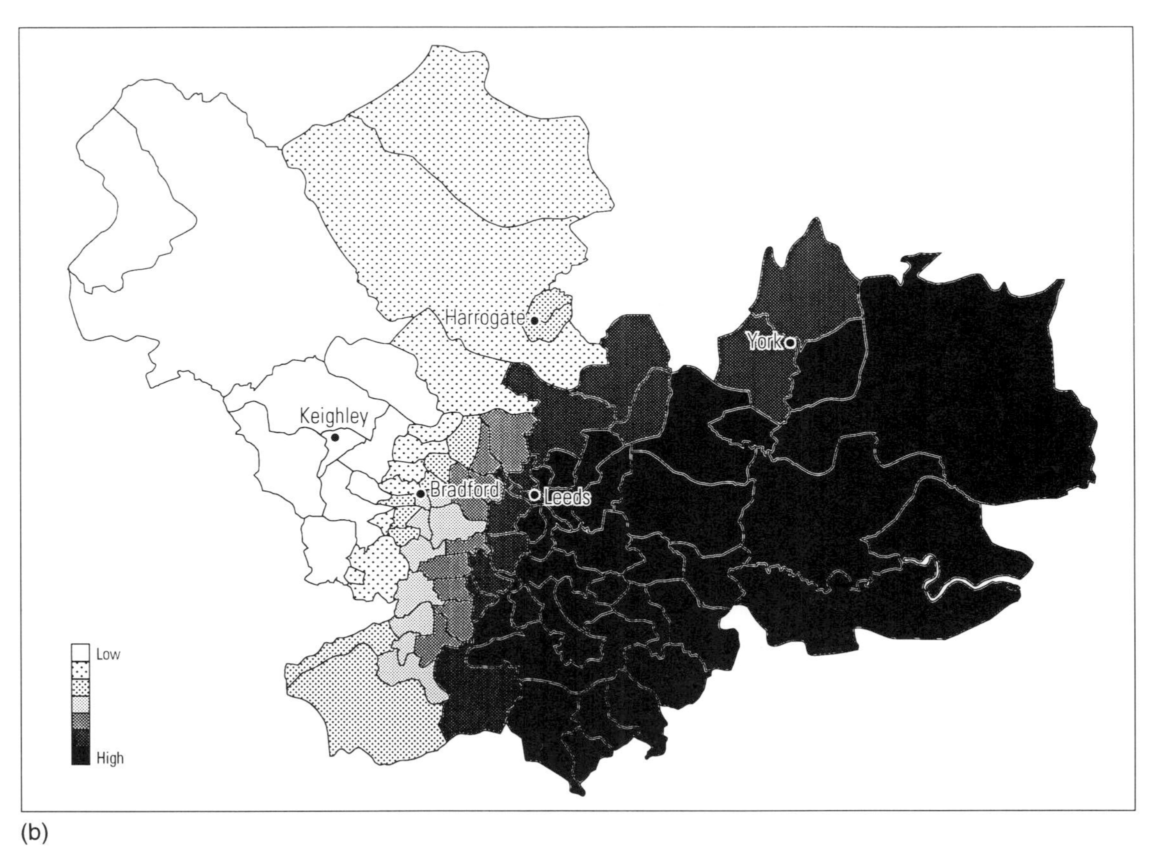

(b)

Figure 7.4 (a) Baseline provision ratio; (b) change in provision ratio

where W_j is the attractiveness of centre j, and Z_j the floorspace available at that centre.

In this situation, as the α parameter is increased, more people will visit the larger centres. In this case, the effective delivery of services to residents of areas close to large centres must be reduced. This is counter-intuitive in that the accessibility to high quality services is high in this situation. Even more perversely, if the floorspace at a large centre rises, then the effective delivery of services to residential areas close to that centre is also likely to fall.

A simple example of this phenomenon is provided in Figure 7.4. Here we have assumed that one of the medium-sized outlying centres – Keighley – is relatively attractive to shoppers. The provision ratios around Keighley are unexceptional; indeed the pattern of service provision across the whole region is a rather balanced one, as shown in Figure 7.4(a). However, if we add another retail outlet to Keighley then we observe changes in the provision ratio which are exactly in accord with expectations (see Figure 7.4(b)). Provision ratios to the northwest of the region are all *reduced* as more customers are pulled into the attractive but small centre at Keighley. This makes other centres less congested; hence provision ratios increase more the further away from Keighley we go!

At one level, one could argue that this is just one indicator among many that need to be balanced against one another in interpretation. However, the point is that effective delivery appears to be measuring something different here to what is measured by the same indicator in public sector applications. What we want to identify are areas that are poorly provided for, preferably by looking at simple maps of the distribution of such an indicator.

Consumer surplus is one possibility here that balances quality of service provision against the accessibility costs of getting it. However, this indicator is particularly difficult to interpret. A new indicator we might define is simply the average size of the centre visited:

$$\mathrm{AS}_i = \frac{\Sigma_j S_{ij} W_j}{S_{i*}} \tag{7.5}$$

We could extend this to become a measure of quality:

$$\mathrm{AS}_i = \frac{\Sigma_{jk} S_{ij} W_j^k r^k}{S_{i*}} \tag{7.6}$$

as a gross measure, or

$$\mathrm{AS}_i = \frac{\Sigma_j (S_{ij}/S_{i*})\ \Sigma_k r^k W_j^k}{\Sigma_k W_j^k} \tag{7.7}$$

as an average measure. r^k is average sales per square foot of competitor k (or some other appropriate quality measure).

7.3.2 Facility-based indicators

In the cases examined in Section 7.3.1, the analysis of residential areas is of interest to the decision-maker (the retail planner). For example, in relation to effective delivery, the identification of areas which are currently poorly provided for will be suggestive of good sites for potential development.

However, what the retail planner really wants to know is how he or she can use this information at a centre level, since this is the sense in which the planner has control over the system. Specifically, which centre(s) have greatest potential for the development of new stores?

The dichotomy of market share against market penetration is again an instructive one here. It is quite possible for a retailer to have a high market share in a centre around which market penetrations are low (e.g. in a suburban centre such as Horsforth); and indeed to have a zero market share in centres around which penetrations are high (e.g. in satellite centres around towns which have a good network of stores, such as Shipley). So what we need are model-based centre indicators which are averages of residential indicators. In the case of market shares we would want to look at market penetrations in the catchment of a centre,

$$\frac{\Sigma_i m_i S_{ij}}{D_j} \tag{7.8}$$

(where m_i is market share), but also, for example, at the quality of service available to the residents of a catchment rather than the quality of services in the centre itself. (Note that this is a logical extension of the argument of

Table 7.4 Market penetration in the neighbourhood of West Yorkshire centres

		Competitor 1	*Competitor 2*	*Competitor 3*	
1	Batley	30.5409	44.0040	25.4552	*
2	Dewsbury	27.5398	46.2497	26.2106	*
3	Pontefract	32.1004	51.9385	15.9612	
4	Castleford	32.0581	45.4559	22.4860	
5	Wakefield	23.9443	51.2152	24.8405	*
6	Wetherby	30.8379	58.6877	10.4744	
7	Ilkley	29.6701	38.4487	31.8811	*
8	Otley	29.3099	39.4875	31.2026	*
9	Horsforth	30.9159	43.5275	25.5566	
10	Pudsey	31.6835	43.1741	25.1423	
11	Leeds	26.3545	47.7613	25.8842	*
12	Skipton	48.4017	47.9836	3.6146	
13	Shipley	34.6642	41.5959	23.7399	
14	Keighley	35.0090	54.2676	10.7234	
15	Bradford	31.8132	42.9769	25.2099	*

Note: An asterisk indicates the presence of a store in this centre for competitor 1.

Clarke and Wilson in Chapter 2 that zone-based indicators *per se* are relatively uninteresting compared with systemic indicators, and provides us with the means by which these systemic indicators can be dissected at the centre level.) Table 7.4 shows market penetrations around the different centres for our three imaginary competitors (and note how again this highlights the relatively balanced pattern of penetration for competitor 1 *vis-à-vis* the other two).

7.3.3 Indicators of model performance

In Section 7.1 we saw that 'group performance' is probably the key indicator of model performance. Group performance can be defined algebraically as

$$\mathrm{GP}_{ij}^{k} = \frac{\Sigma_i S_{ij}^{k}}{D_j^{k'}} \tag{7.9}$$

where S_{ij}^{k} is the model estimate of flows of customers to type k stores in j; $D_j^{k'}$ is the known revenue of the store.

In assessing the strength of the model, the mean value of group performance will tell us little, of course, as this will tend to unity in most cases. The standard deviation or variance may be of more interest, as well as standard goodness-of-fit measures such as chi-squared. In practice, however, confidence in the performance of the model is more likely to be achieved through intuitive and relatively informal measures. An example is a version of the '80–20 rule' in which 80% of predictions must be within ±20% of observed sales, although the precise bounds of acceptability may often be tighter. The best way to view this information may be simply to view group performance as a frequency distribution.

However, there are also problems associated with the use of group performance as an indicator of the 'goodness of fit' of a model. One problem is that, where there is a mismatch between revenue predictions and the observed data, it is often difficult to distinguish between imperfections in the modelling procedure and the existence of some real factors which are causing particular stores to perform poorly or well. Of course these two elements are often interconnected. For example, if an appraisal of the performance of a chain of stores reveals that takings are generally lower than expected in all cases when the store is on two floors rather than one, then perhaps the model should be respecified in the light of these findings. However, when the performance of car dealers is compared, it may be that in cases where group performance is high, the dealer is simply better at selling cars than his poorly performing colleagues.

Another problem with group performance, or appropriate aggregate measures based on it, is that it is a one-dimensional index based only on performance at facilities. Similar calibration statistics which are frequently used and suffer from the same drawback are average distance travelled, either across the whole system or from a particular zone or to a particular facility.

Table 7.5 A comparison of observed and predicted entropy measures for West Yorkshire centres

		Predicted		*Observed*	
		Value	*Rank*	*Value*	*Rank*
1	Batley	–2.022153	5	–0.4116846	15
2	Dewsbury	–2.098905	4	–1.494179	7
3	Pontefract	–1.6263	10	–1.015780	12
4	Castleford	–1.692726	9	–1.421638	8
5	Wakefield	–1.916489	6	–1.806956	3
6	Wetherby	–1.050453	13	–1.644373	4
7	Ilkley	–0.9936571	14	–0.4355199	14
8	Otley	–1.316977	12	–1.116124	11
9	Horsforth	–1.842435	7	–1.243957	10
10	Pudsey	–1.824376	8	–1.330698	9
11	Leeds	–2.350268	2	–1.513040	6
15	Bradford	–2.508019	1	–2.952129	1

So all these measures tell us nothing about the pattern of flows within our models. In an extreme case, we might model flows to a facility simply in proportion to the provision of floorspace, and this might provide an adequate predictor of revenue generated but, in all probability, will produce a hopeless representation of the associated flow patterns. One way of representing the two-dimensional flow patterns is through an entropy-based measure which reflects the level of concentration in the pattern:

$$\sum_{i} s_{ij} \log s_{ij} \tag{7.10}$$

where

$$s_{ij} = S_{ij}/S_{i*} \tag{7.11}$$

If information about the actual flow patterns is known (either partially or completely) then we may compare the observed equivalent:

$$\sum_{i} s'_{ij} \log s'_{ij} \tag{7.12}$$

Both observed and predicted entropy measures for West Yorkshire centres are shown in Table 7.5. In general, the values in the table are lower for the predicted entropy, indicating a greater dispersion of flows. This is expected, as the observed patterns are based on a relatively sparse sample of real flows. When we compare the ranking of dispersion by centre, it appears that Batley, Wetherby and Keighley might be investigated further to look for irregularities in either the observed or the predicted patterns.

Note that commonly used regression-based measures (of s_{ij} as the dependent variable against s'_{ij} as the independent variable) may also be useful here, although we can expect very high correlations in practice for all but the poorest models. Also, with such regression models it is hard to cope with zero flows (which will be numerous) especially where either the observed or the predicted flow is zero and the other is non-zero (see Williams and Fotheringham (1984) or Knudsen and Fotheringham (1986) for a fuller discussion of the strengths and weaknesses of this and alternative goodness-of-fit measures).

In all of this we need to bear in mind the literal definition of performance indicators as *indicators* of (model) performance. What this means in the present context is that we are concerned less with statistical verification than with establishing intuitively appealing measures which give confidence, or alert us to potential danger if something is out of place. One of the simplest but most effective comparisons of flow structures may be to rank the observed and predicted flows to a centre and compare them using a rank correlation coefficient such as Spearman's.

Finally, the golden rule about 'batteries of performance indicators' (cf. Chapters 2, 5 and Birkin and Clarke 1987) applies. If at all possible, produce all the different kinds of indicators, so that if something is amiss it is more likely to show up as an oddity on at least one of the indicators.

8 Applications of performance indicators in urban modelling: subsystems framework

M. Birkin, G.P. Clarke, M. Clarke and A.G. Wilson

8.1 INTRODUCTION

The aim of this chapter is to review a range of urban modelling applications involving the use of the performance indicator framework outlined in Chapters 3–7. The examples are drawn from the UK; Chapter 9 examines indicators in an Italian context. We begin with the crucial variable, income, since this is central to the contemporary social geography of any city or region. This is followed by examples from our work on local labour markets and retailing. The examples are partial and illustrative at this stage; articulating the full performance indicator framework identified in Chapters 3–7 awaits a further volume!

8.2 INCOMES

Although regional incomes and expenditure are available for different household types in the UK (through the published household surveys 'Family Expenditure Survey' and 'New Earnings Survey') no published statistics are available for income within individual cities or areas within cities. In order to derive household incomes at the small area level we need to calculate the age (a), sex (s), occupation (b) and industry (g) breakdown of all economically active persons in a particular spatial zone (i); i.e. the array $\{P_i^{bags}\}$. (We shall return to income from other sources later.) Again, this full array is not available from census reports but can be estimated using *micro-simulation techniques*. The argument here follows Birkin and Clarke (1988). The modelling exercise is to calculate $\{P_i^{bags}\}$ given *known* values of the following arrays:

$\{P_i^{as}\}$ (UK SAS Census Table 26)
$\{P_i^{asg}\}$ (UK SAS Census Table 46)
$\{P_i^{ab}\}$ (UK SAS Census Table 50)
$\{P_i^{bg}\}$ (UK SAS Census Table 44)

where SAS refers to 'Small Area Statistics'.

The two most common methods of solving this exercise are balancing factor methods (cf. Wilson 1970) and iterative proportional fitting (IPF) (cf. Fienberg 1970). The latter is the fundamental procedure for building synthetic micro-data sets, estimating full conditional probability distributions from partial data irrespective of the number of attributes involved. In this case IPF routines can be used to calculate $\{P_i^{bags}\}$ by sampling for industry and occupation on the basis of known age, sex and location (see Birkin and Clarke 1988; Birkin 1987). Having obtained estimates of the array $\{P_i^{bags}\}$ we must now convert the raw population totals into estimates of household income. That is,

$$I_i^{bags} = P_i^{bags} I_r^{bags} \tag{8.1}$$

where r refers to regional income figures available from published sources. To calculate the full array $\{I_r^{bags}\}$, however, does require further IPF calculations. The modelling exercise is to calculate $\{I_r^{bags}\}$ given *known* values of the following arrays:

earnings by occupation and sex	$\{I^{bs}\}$	(New Earnings Survey Tables 122, 123)
earnings by industry and sex	$\{I^{gs}\}$	(New Earnings Survey Tables 104, 105)
earnings by age, sex and region	$\{I_r^{as}\}$	(New Earnings Survey Tables 128)

The results of this exercise can be seen on a variety of spatial scales (see Birkin and Clarke 1989).

To obtain total household incomes in a particular area we also need to estimate the number of households dependent upon state benefits (primarily the aged and those looking for work): i.e.

$$I_i = \sum_{bags} I_i^{bags} + \sum_{0} I_i^0 \tag{8.2}$$

where I_i^0 are households with no earned income. The calculation of households dependent on state benefits is best illustrated through an example – the census ward of Armley in Leeds (see Table 8.1). If we define state benefits as falling into two categories, 'state pensions' and 'employment benefit', then we clearly need to establish the number of households containing pensioners and those containing unemployed people.

We can estimate the number of households with no economically active residents in employment by applying the basic employment rate to each economically active individual. Hence in Armley the number of households with no wage earners is calculated as (from Table 8.1)

$$(2{,}673) \times (0.091) + (3{,}444) \times (0.091) \times (0.091) = 272$$

For households containing pensioners, we can identify a rate of dependence on state benefits from the Family Expenditure Survey (Table 8.2). Specifically, Table 8.2 shows that 60.6 per cent of households containing two or more pensioners are completely dependent on state benefits.

Table 8.1 Estimated household structure, Armley, 1990

		All households	*Households with pensioners only*
Households with no economically active residents	1 person	1,563	1,408
	2+ persons	893	788
Households with economically active residents	1 economically active person	2,673	
	2+ economically active persons	3,444	

Source: Updated figures from Birkin and Clarke 1988

Table 8.2 Dependency in Armley, 1990

Household type		*Number*	*Dependency rate*	*Dependent households*
Pensioner – one adult		1,408	0.606	853
Pensioner – two+ adults		788	0.420	331
Unemployed		272	1.000	272
Total dependent households	1,456			
Total households	8,573			
Dependency rate	1,456/8,573 = 17.0%			

To calculate the proportion of households partially dependent on state benefits, we add together all households containing pensioners and all households containing unemployed persons from Table 8.2:

$$1{,}408 + 788 + (2{,}673 \times 0.091) + [3{,}444 \times (1 - 0.909)] + [3{,}444 \times 1 - (0.909 \times 0.909)] = 3{,}369$$

$$\text{dependency ratio} = 3{,}369/8{,}573 = 0.393$$

Combining the total number of households wholly and partially dependent on state benefits (see Table 8.3 for different geographical scales) with household estimates of earned income provides us with estimates of total household and regional income figures $\{I_i^h\}$ (Table 8.4).

Table 8.3 Households dependent on state benefits, 1990

		Households wholly dependent on state benefits	*Households partially dependent on state benefits*
Wards of Leeds			
1	Armley	17.1	39.3
2	Burmantofts	20.6	47.7
3	Chapel Allerton	18.0	44.7
4	City & Holbeck	25.0	54.1
5	Harehills	19.7	46.7
6	Headingley	17.5	42.5
7	Hunslet	17.5	42.1
8	Middleton	15.9	38.6
9	Richmond Hill	18.6	44.1
10	Seacroft	17.9	43.7
11	University	24.1	56.5
12	Kirkstall	19.9	44.1
12	Inner City	19.4	45.5
13	Leeds M.D.	16.0	36.3
1	West Yorkshire	15.6	35.3
2	Yorks + Hum	15.9	36.2
3	UK	14.8	32.8
4	EEC	13.9	31.0
1	Manchester	16.0	36.5
2	Liverpool	17.2	41.6
3	Doncaster	14.7	36.3
4	Sheffield	17.3	39.3
5	Newcastle	16.4	38.9
6	Birmingham	14.5	33.9
7	Bradford	15.7	35.9
8	Calderdale	16.8	36.4
9	Huddersfield	15.9	34.6
10	Leeds	15.8	35.5
11	Wakefield	14.7	34.7
12	Hull	16.2	37.2
13	Nottingham	14.7	33.8
14	Glasgow	16.3	38.4

Table 8.4 Incomes, 1990

Inner City wards of Leeds		*Gross income*	*Income per head*	*Income/capital*
1	Armley	2,298.1	268.6	104.2
2	Burmantofts	1,998.4	233.3	91.6
3	Chapel Allerton	2,425.5	294.2	107.4
4	City & Holbeck	1,715.7	228.8	96.7
5	Harehills	2,143.6	241.0	87.9
6	Headingley	1,992.3	308.2	135.8
7	Hunslet	1,548.7	244.7	96.2
8	Middleton	2,002.7	262.2	94.2
9	Richmond Hill	1,889.5	239.8	90.1
10	Seacroft	1,931.6	241.0	89.5
11	University	1,618.4	241.7	108.2
12	Kirkstall	2,137.3	261.7	134.5
13	Inner City	23,701.4	254.4	103.0
Other regions				
1	West Yorkshire	239.6	315.7	117.9
2	Yorks + Hum	554.9	315.0	117.2
3	UK	6,863.6	349.0	128.9
4	EEC	38,997.8	338.6	120.5
1	Manchester	179.2	328.3	122.3
2	Liverpool	115.3	321.6	113.3
3	Doncaster	32.0	336.1	121.0
4	Sheffield	70.8	309.5	118.5
5	Newcastle	83.6	286.8	109.3
6	Birmingham	178.8	331.9	118.5
7	Bradford	54.5	313.6	114.4
8	Calderdale	23.1	318.5	121.5
9	Huddersfield	27.5	316.6	119.1
10	Leeds	79.3	305.9	116.8
11	Wakefield	33.1	339.9	123.4
12	Hull	50.0	316.9	116.6
13	Nottingham	92.9	340.0	127.1
14	Glasgow	138.6	312.6	111.0

Note: Income is measured in thousands of pounds for Inner City wards, and in millions for the other regions.

8.3 WORKPLACE INDICATORS

It was argued in Chapter 5 that a crucial array for measuring labour market performance was $\{T_{ij}^{wbg}\}$ or, by substituting age (a) and sex (s) for person type (w), $\{T_{ij}^{bags}\}$. This array is clearly the population groups we examined in Section 8.2 with additional origin–destination labels which define the full interaction set between individual workers and specific jobs throughout the city. As we saw in Section 8.2 it is possible to calculate $\{P_i^{bags}\}$ through IPF routines. At the supply end, employment data are available in Britain for the arrays $\{Z_j^{bg}\}$ and $\{Z_j^{gs}\}$, where Z_j is the number of jobs available in zone j, at levels of resolution down to the individual census ward (or indeed postal sector). Interaction data between wards are available, disaggregated by mode of travel (m) and sex, i.e. $\{T_{ij}^{ms}\}$. In mathematical programming terms we thus have the following problem:

$$\min_{\{T_{ij}^{bags}\}} \sum_{bag} T_{ij}^{bags} - \sum_{m} \hat{T}_{ij}^{bags} \tag{8.3}$$

subject to

$$\sum_{j} T_{ij}^{bags} = \hat{P}_i^{bags} \tag{8.4}$$

$$\sum_{iag} T_{ij}^{bags} = \hat{Z}_j^{bs} \tag{8.5}$$

$$\sum_{iba} T_{ij}^{bags} = \hat{Z}_j^{gs} \tag{8.6}$$

where a circumflex indicates published data available. Both IPF or spatial interaction routines could be used to solve this problem (see Birkin and Clarke 1987).

Once the array $\{T_{ij}^{bags}\}$ has been established it is a fairly straightforward procedure to generate and sample from cumulative probability distributions that will assign individuals to a zone of workplace.

$$p(j\colon i, b, a, g, s) = T_{ij}^{bags} \Big/ \sum_{ibgsa} T_{ij}^{bags} \tag{8.7}$$

Having estimated the crucial array $\{T_{ij}^{bags}\}$ we can now offer far more detail on the operations and interdependences of the local labour market than can be gleaned from secondary sources. First we look at a variety of performance indicators based on the array $\{T_{ij}^{bags}\}$ and then we explore the powerful use of the array in examining changes in the local labour market.

8.3.1 Residence-based indicators

We can illustrate the variations in performance indicator figures by concentrating on three very different areas of the city of Leeds in the UK: 'North' is an affluent northern suburb of the city characterized by many higher income households and few locally available jobs; 'Richmond Hill' is an inner city, traditional manufacturing area with lower income households; and 'Harehills' is another inner city location but one containing a large residential population and few locally available jobs.

The main emphasis for residence-based indicators is on different person types, disaggregated by social class, age and sex. We believe that this type of analysis makes an important contribution to the 'personal characteristics' literature, which has already made inroads into how (un)employment (and deprivation more generally) varies with attributes such as age, sex, social class, tenure, race, qualifications etc. (Van Dijk and Folmer 1985; Gordon 1986, Table 4; Worrall 1986; Hasluck 1987).

Employment and unemployment rates

The first indicators simply present disaggregated basic accounting totals: the probability of residents in particular areas being employed or unemployed respectively.

Table 8.5 shows the very high rates of unemployment and low rates of employment probabilities for the manual workers, especially in the younger age groups, in both Harehills and Richmond Hill. North, however, in the more affluent leafy suburbs, tends to have much higher employment probabilities and a lower rate of unemployment for all social classes. These kinds of figures support earlier research on activity patterns such as that of Webber (1978) who discovered a similar probability of the 'best' areas having much lower levels of unemployment than other areas, even for the unskilled categories.

Also striking from Table 8.5 are the very much lower rates of unemployment for female workers except for the professional group (note that the zeros in Richmond Hill indicate none present rather than nil unemployment). These types of result emphasize the benefits of the simulation approach, combining information on age, sex, occupation and location. Whilst there are numerous studies on the spatial or occupational variations in (fe)male unemployment for example, there are few which can effectively combine all the major personal attributes.

Average distance travelled

The indicator here is written as

$$C_i^{bags} = \frac{\Sigma_j T_{ij}^{bags} d_{ij}}{\Sigma_j T_{ij}^{bags}} \tag{8.8}$$

where d_{ij} is the actual distance (or time taken) of the journey to work.

The obvious feature of Table 8.6 is the much longer average distance travelled from all residents in North. This, of course, may not indicate a

Table 8.5 Employment and unemployment rates (percentages)

	Professional and managers		*Semi-skilled and unskilled*	
	Male	*Female*	*Male*	*Female*
Employment rates				
North				
16–24	96.10	93.60	82.2	90.4
25–44	97.40	98.80	89.8	96.2
45+	97.20	98.70	89.6	95.9
Richmond Hill				
16–24	89.1	100	55.9	97.2
25–44	88.6	100	64.7	98.7
45+	92.8	100	75.9	98.2
Harehills				
16–24	96.0	88.5	59.3	88.5
25–44	97.8	96.0	67.5	97.1
45+	98.2	94.2	58.6	96.8
Unemployment rates				
North				
16–24	3.90	6.40	17.80	9.60
25–44	2.60	1.20	10.20	3.80
45+	2.80	1.30	10.40	4.10
Richmond Hill				
16–24	10.90	0.00	44.10	2.80
25–44	11.40	0.00	35.30	1.30
45+	7.20	0.00	24.10	1.80
Harehills				
16–24	4.00	11.50	40.70	11.50
25–44	2.30	4.00	32.50	2.90
45+	1.80	5.80	41.40	3.20

Table 8.6 Average distance travelled (km)

	Professional and managers		*Semi-skilled and unskilled*	
	Male	*Female*	*Male*	*Female*
North				
16–24	7.23	8.20	7.87	3.67
25–44	7.45	9.21	5.54	3.85
45+	8.52	3.90	5.63	5.18
Richmond Hill				
16–24	3.87	2.78	4.82	4.13
25–44	2.60	4.68	5.12	4.45
45+	3.97	3.90	4.14	3.75
Harehills				
16–24	3.75	1.50	4.14	2.75
25–44	2.82	3.05	3.80	3.50
45+	2.60	2.30	3.86	3.85

problem where the professional group is concerned since journey time may simply have been traded for housing and environmental quality. The high figures for the unskilled groups are perhaps of more concern, not only in North but also in Harehills and Richmond Hill. Both these wards have longer average travel distances to work for unskilled groups than professional groups, who seem to work in the locality itself or in the city centre nearby. (We can shed more light on this area by looking at the number of residents finding jobs in their own localities – the degree of self-containment – in the next section.)

It is interesting that rates of female travel in the local labour market are roughly in the same proportion as male rates, with the exception of professionals in North. This is generally because there are fewer jobs for professional women outside the central areas and hints at the low levels of female penetration in managerial posts in more industrial areas (see also the workplace indicators).

Degree of self-containment

The indicator to be explored here takes the following form:

$$X_i^{bags} = \frac{T_{ii}^{bags}}{Z_i^{bags}} \tag{8.9}$$

where Z_i is the number of jobs available in zone i and clearly measures the degree to which residents can find jobs in their own locality.

As one might expect given the different numbers of jobs in each ward, there is quite a wide variation between social groups, age groups and locations (Table 8.7). In North, there are few male or female professional and managerial jobs and most of this group are forced to commute (remember the long journey times of the previous indicator). The degree of self-containment in Harehills is very much higher, although there are very many fewer professional workers demanding jobs from Harehills. The same is true of Richmond Hill although in this case there also seems to be a greater degree of out-commuting. The anomaly here is the 25–44 age group for male workers. It is difficult to conclude at this stage whether this is simply a small numbers problem or a reflection of the larger number of middle-aged managerial posts associated with local manufacturing industry.

On the unskilled side the figures are generally higher showing that some jobs are available locally to these areas (although bear in mind that the average distance travelled for these groups was quite high, suggesting relatively long-distance commuting for those who cannot find jobs locally). Harehills is the exception here, offering very little in the way of local employment for the low skilled and unskilled.

Table 8.7 Degree of self-containment (residents)

	Professional and managers		*Semi-skilled and unskilled*	
	Male	*Female*	*Male*	*Female*
North				
16–24	0.175	0.182	0.429	0.745
25–44	0.115	0.105	0.488	0.696
45+	0.117	0.278	0.410	0.619
Richmond Hill				
16–24	0.154	0.40	0.213	0.152
25–44	0.684	0.182	0.194	0.216
45+	0.158	0.111	0.305	0.243
Harehills				
16–24	0.3	0.80	0.04	0.15
25–44	0.40	0.308	0.137	0.147
45+	0.38	0.60	0.085	0.171

8.3.2 Facility based indicators

Degree of market share

This indicator measures the number of jobs (in industry type g) provided in a particular workplace zone as a percentage of the total number of jobs across the whole market. Formally it is written as

$$X_j^{bags} = \frac{Z_j^{bags}}{\Sigma_i Z_i^{bags}} \tag{8.10}$$

To explore this indicator we concentrate on those areas that have the highest overall employment totals: the inner wards of Leeds.

It is evident from Table 8.8, part (a), that the central area of the city (including the city centre itself and the industrial area of Holbeck) still dominates the local labour market, providing over a third of all male professional jobs, half the total female professional jobs and a third of all the male and female unskilled manual jobs. If we disaggregate to male manufacturing employment (Table 8.8, part (b)) we can see the high rates of low skilled employment concentrated in the south of the city. (Although we have argued that Richmond Hill is a fairly large employment zone it nevertheless accounts for less than 10 per cent of jobs for any social class/age category.)

Size of market area

In order to determine the geographical extent of the market area for each ward we can calculate average distance travelled by persons working in the ward

Table 8.8 Degree of market share

(a) City & Holbeck

	Professional and managers		*Junior non-manager*			*Unskilled*
	Male	*Female*	*Male*	*Female*	*Male*	*Female*
16–24	0.389	0.436	0.447	0.42	0.397	0.399
25–44	0.391	0.586	0.416	0.474	0.444	0.305
45+	0.400	0.509	0.424	0.376	0.384	0.356

(b) Manufacturing (male)

	Skilled manual	*Semi-skilled manual*	*Unskilled*
City & Holbeck			
16–24	0.151	0.192	0.185
25–44	0.157	0.173	0.227
45+	0.187	0.171	0.207
Hunslet			
16–24	0.130	0.153	0.171
25–44	0.113	0.152	0.170
45+	0.119	0.148	0.138
Richmond Hill			
16–24	0.057	0.03	0.085
25–44	0.06	0.035	0.068
45+	0.05	0.04	0.03

(the opposite to the residential 'average distance travelled' indicator). This is written as

$$XX_j^{bags} = \frac{\Sigma_i T_{ij}^{bags} d_{ij}}{\Sigma_i T_{ij}^{bags}} \tag{8.11}$$

The obvious feature of Table 8.9 is the longer average distance travelled by professional workers coming into the workplace zones (compare with the distances travelled from the residential areas in Table 8.6) and the corresponding decline in distances travelled as one moves through the social classes. The generally high figures compared with the residential totals suggest a considerable amount of cross-boundary flows between areas and that even areas such as Hunslet and Richmond Hill attract professional and skilled workers from outer suburbs. We shall look in more detail at the interactions themselves in the next section.

Degree of self-containment

The third workplace indicator is simply the reverse of the last residence-based indicator. It measures the degree to which a particular workplace zone employs residents from within that zone (aggregated over all job types). Formally the indicator is expressed as

Table 8.9 Geographical size of market area (male)

	Professional	*Skilled/man*	*Semi-skilled*	*Unskilled*
City				
16–24	7.25	5.35	5.31	3.03
25–44	8.23	5.80	5.73	4.45
45+	8.99	5.83	4.49	4.40
Hunslet				
16–24	7.81	5.16	5.29	4.26
25–44	9.65	5.53	5.04	5.29
45+	7.34	5.67	5.07	4.49
Richmond Hill				
16–24	5.24	4.28	5.50	5.81
25–44	9.19	5.09	5.33	4.62
45+	7.14	5.20	5.31	5.48

$$Y_j^{bags} = \frac{T_{jj}^{bags}}{Z_j^{bags}} \tag{8.12}$$

For this indicator we return to our three previous study zones, North, Harehills and Richmond Hill (Table 8.10).

We saw in the last section the openness of workplace areas that have a large number of jobs. In comparison we see that North and Harehills, which have comparatively fewer jobs at the extreme ends of the labour market, have a much greater degree of self-containment. The interesting ward again here, however, is Richmond Hill. Although it can obtain a third to a half of its

Table 8.10 Degree of self-containment (workplace)

	Professional and managers		*Semi-skilled and unskilled*	
	Male	*Female*	*Male*	*Female*
North				
16–24	1.00	1.00	1.00	0.709
25–44	0.915	1.00	0.84	0.728
45+	0.875	1.00	0.862	0.904
Richmond Hill				
16–24	0.111	0.29	0.374	0.332
25–44	0.146	0.265	0.270	0.307
45+	0.069	0.250	0.482	0.263
Harehills				
16–24	0.607	1.00	1.00	0.603
25–44	0.405	0.50	0.94	0.629
45+	0.687	0.75	1.00	0.70

unskilled labour force locally, it is very open to commuting from other areas of unskilled labour market nearby, namely Burmantofts, Seacroft, Hunslet and Halton (Figure 8.1). This helps to account for the relatively large average commuting distances seen in the last section. The large concentration of unskilled labour in the south and east of the city is clearly a major feature of the local labour market.

We can get a clearer picture of the degree of interaction for unskilled workers across south and east Leeds by plotting the actual flows into Richmond Hill. Figure 8.2 illustrates the scale of commuting involved.

Similarly we can plot the flows of professional workers into Richmond Hill, which also show both low rates of local penetration (Table 8.10) and long average journeys to work (Table 8.9). These are shown in Figure 8.3.

Summary

Out of a large and powerful array of data $\{T_{ij}^{bags}\}$, we have selected a number of illustrative indicators and wards of the city to examine the degree of spatial interdependence in the local labour market. Although selective in approach, we have already begun to see the degree to which the local labour market is open to commuting and which type of residents (in which types of locality) are most disadvantaged in terms of job accessibility. Specifying a fuller range of attributes (tenure, race, educational attainment, job skills etc.) would clearly make the analysis even more powerful. This remains an important research task for the future. (For a similar analysis of the market for training see Birkin and Clarke 1992.)

8.4 CHANGING THE LABOUR MARKET INDICATORS

We have argued that a combination of the micro-simulation procedure and more conventional spatial interaction modelling has enabled us to analyse the important spatial interdependences within local labour markets. This has helped to take us beyond the more traditional indicators of zonal employment and unemployment. However, we are aware that the system as described is currently static. In reality, the relationships between people and workplaces are changing constantly. Although people enter and leave the labour force and migrate between locations and occupations at frequent intervals the most dramatic impacts occur when there are large-scale redundancies or new job creation. In Birkin *et al.* (1990) we offered the simple example of the effects of adding 100 new professional or 100 unskilled jobs into an inner city area of Leeds with high unemployment levels, exploring the likely catchment areas and the likely impacts on the local unemployment rates. In the rest of this section we explore the more realistic scenario of a new science park being located in the city and creating 200 new 'high technology' jobs. The possible locations are Kirkstall, the University, Weetwood and Wetherby (Figure 8.4).

In this section our comparison will take two forms. First, we review a

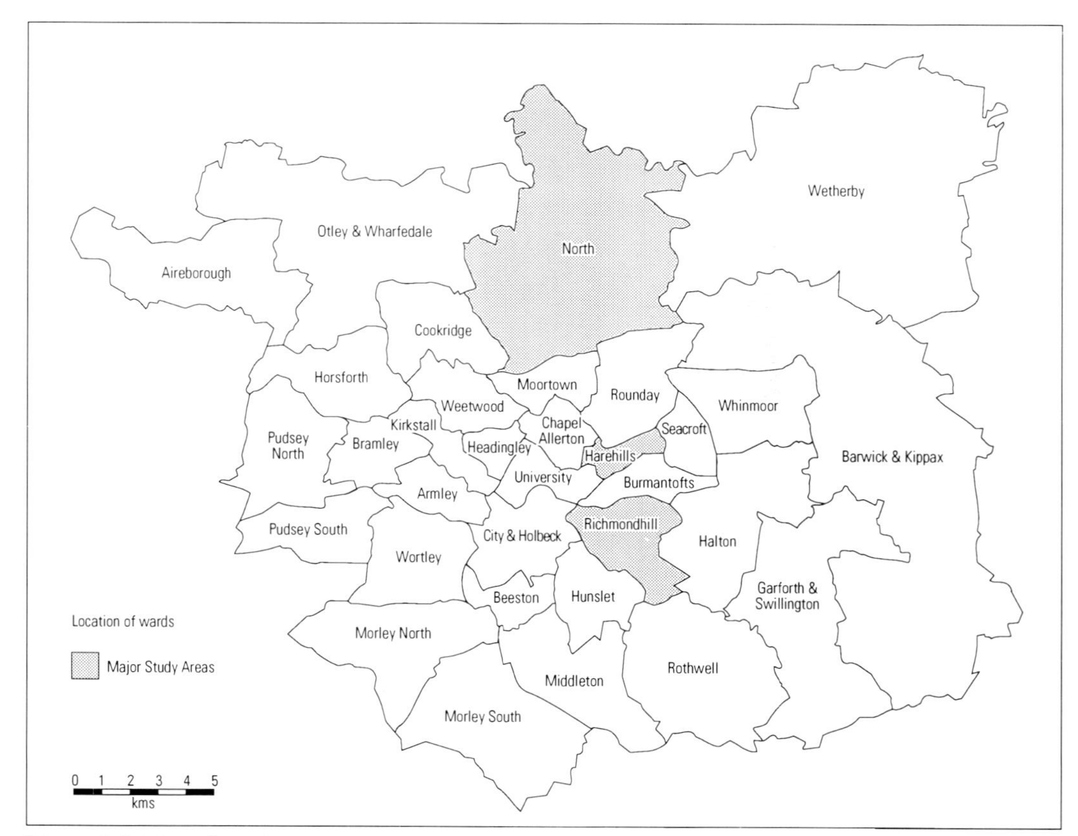

Figure 8.1 Map of Leeds wards

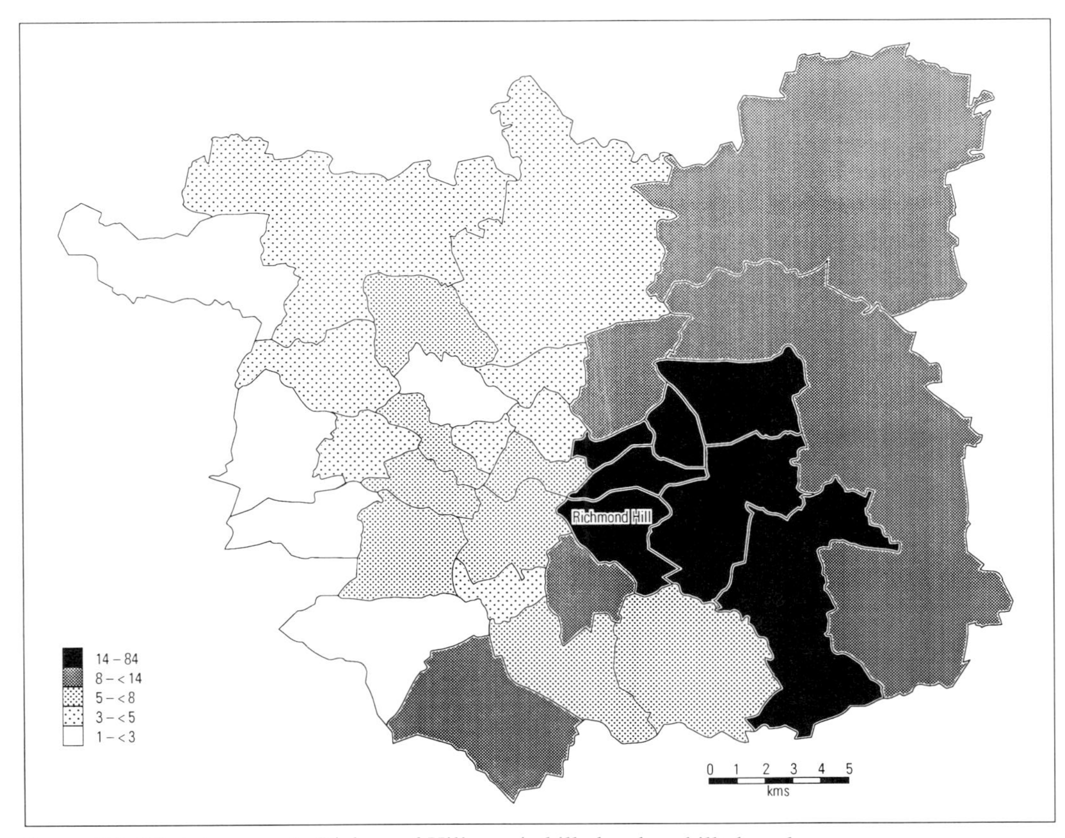

Figure 8.2 Work journeys to Richmond Hill: semi-skilled and unskilled workers

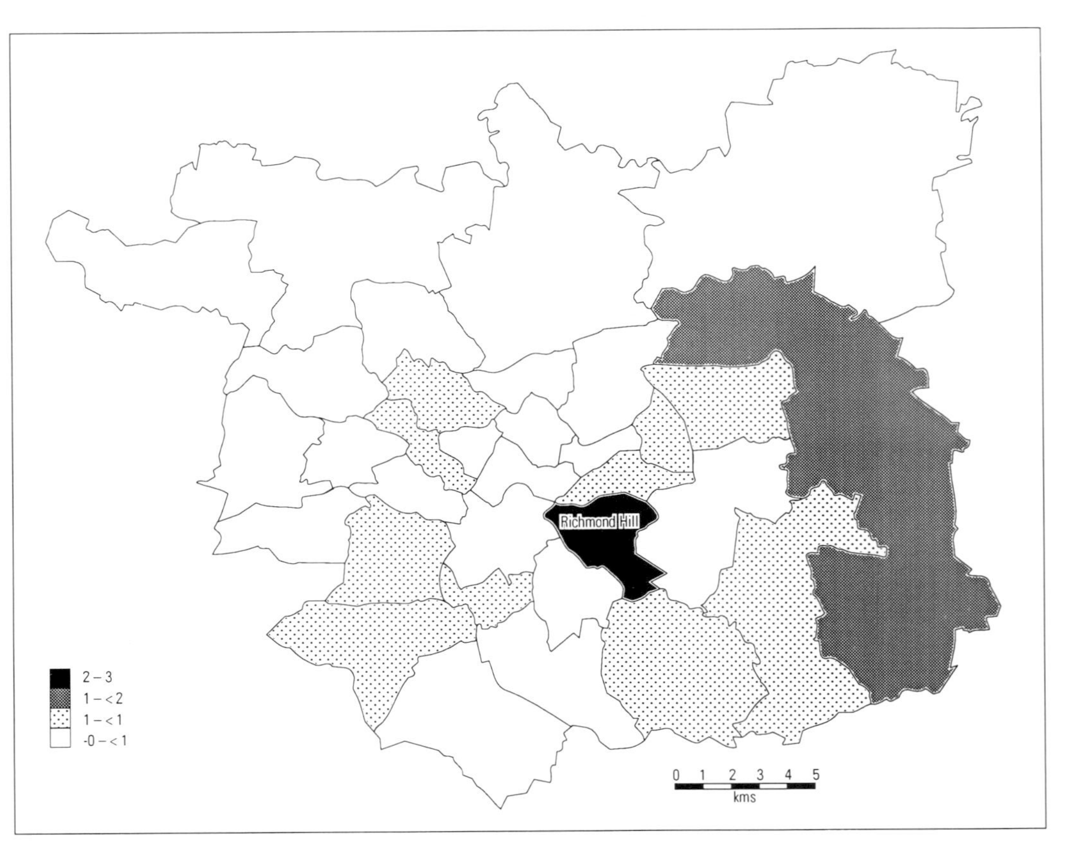

Figure 8.3 Work journeys to Richmond Hill: professional workers

Figure 8.4 Location of four potential sites for a new science park in Leeds (■)

variety of census- and model-based performance indicators for these sites. Second, we conduct a model-based impact analysis to show from where workers would potentially be drawn to these science parks.

Table 8.11 shows that Kirkstall Valley and Leeds University have fairly similar potentials across the whole range of characteristics. The catchment area of Kirkstall is slightly tighter, as defined by the average distance travelled, and Kirkstall also has higher labour market attractiveness, lower unemployment and higher car ownership. Furthermore there are proportionately more graduates and more scientific workers in the labour market area. However, at present there is only a single high technology worker in postal district LS5, against 191 in LS2.

Table 8.11 Comparison of key indicators

	Wetherby (LS22)	*North Leeds (LS16)*	*University (LS2)*	*Kirkstall Valley (LS5)*
Existing high technology employment	626	224	191	1
Average distance travelled (km)	11.2	8.4	8.2	7.9
Labour market attractiveness (i.e. ability to reside locally)	0.12	0.21	0.17	0.20
Unemployment (for professional residents) (%)	2.64	2.96	3.66	3.51
Graduates (in locality) (%)	18	17	13	14
No car households (%)	32	40	48	46
Percentage of workforce classified as professional	4.5	4.0	3.4	3.6

On many of the quality indicators Wetherby appears as the most attractive site, with the lowest unemployment, highest levels of car ownership and largest proportions of both graduate and scientific workers in its labour market. However, measuring the attractiveness of the labour market purely in relation to the observed preferences of high technology workers shows Wetherby to be the least attractive of these locations (this might reflect the fact that high technology workers place a relatively high value on urban amenities not to be found in Wetherby). Wetherby also has the largest existing workforce, and the biggest physical catchment area. On the other hand, the North Leeds site has slightly higher unemployment, fewer graduates and scientific workers, and quite significantly lower levels of car ownership relative to Wetherby. However, it does have the most attractive labour market of them all, with plenty of high quality housing nearby.

On this evidence it is hard to separate Wetherby and North Leeds, which would provide attractive workplace locations for a labour force migrating into the region. Both sites can also draw on a high quality labour force 'already at work' within the catchment. The only obvious advantage of the central sites – the University and Kirkstall Valley – in labour market terms is that higher unemployment levels in the existing catchment may be indicative of a more readily accessible pool of labour.

Finally Figures 8.5 and 8.6 demonstrate the specific postal districts from which one would expect a new science park in Wetherby and Weetwood, respectively, to draw a sample 200 workers in the different cases. As observed above, we can now see that Wetherby has the most spatially extensive

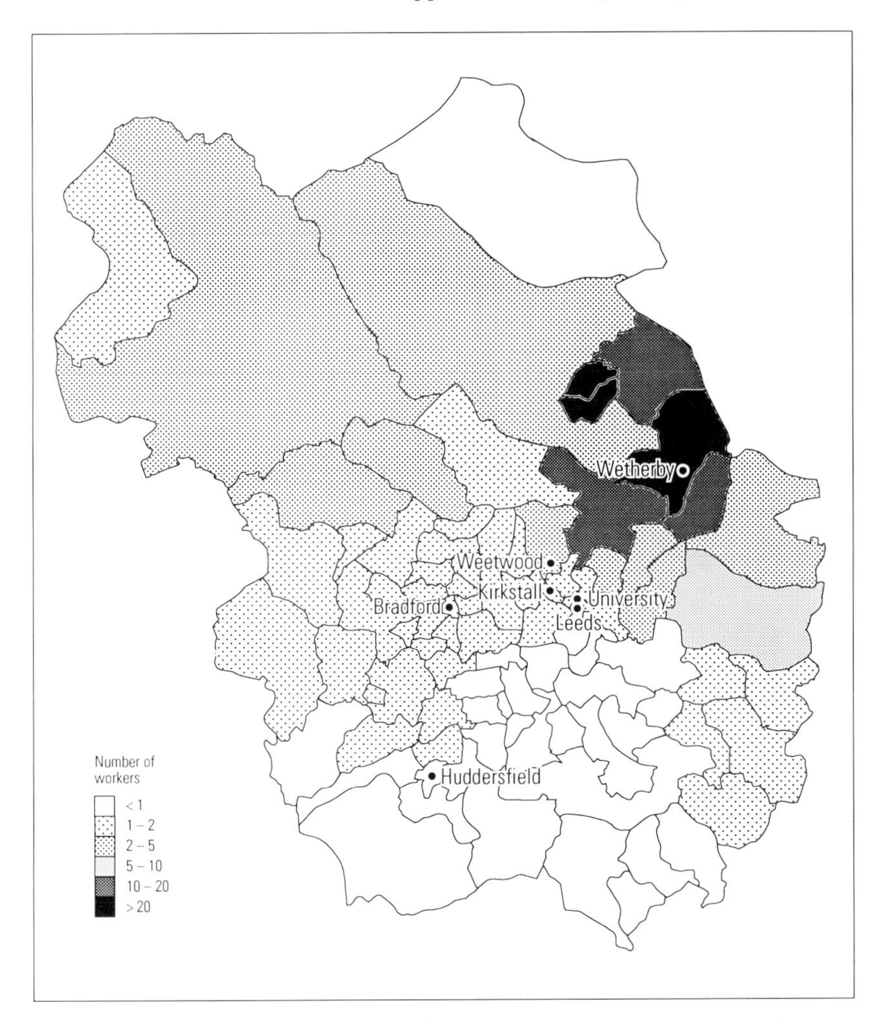

Figure 8.5 The labour market area for 200 workers on a science park in Wetherby

workforce catchment area (although some workers will also be drawn from the York postal area which is off the map to the east – these flows are modelled, but are not mapped here). The Weetwood site draws workers from across a broad area of north-east and north-west Leeds.

One of the central issues such catchment area maps raise concerns accessibility, particularly accessibility to key trunk roads/motorways and good car parking facilities. Added to this must surely be access to work at peak morning and evening times. Any new proposal for a science park must look at the implications for traffic flow and journey time based on such likely journey to work patterns.

The example here is again purely illustrative although taken from current

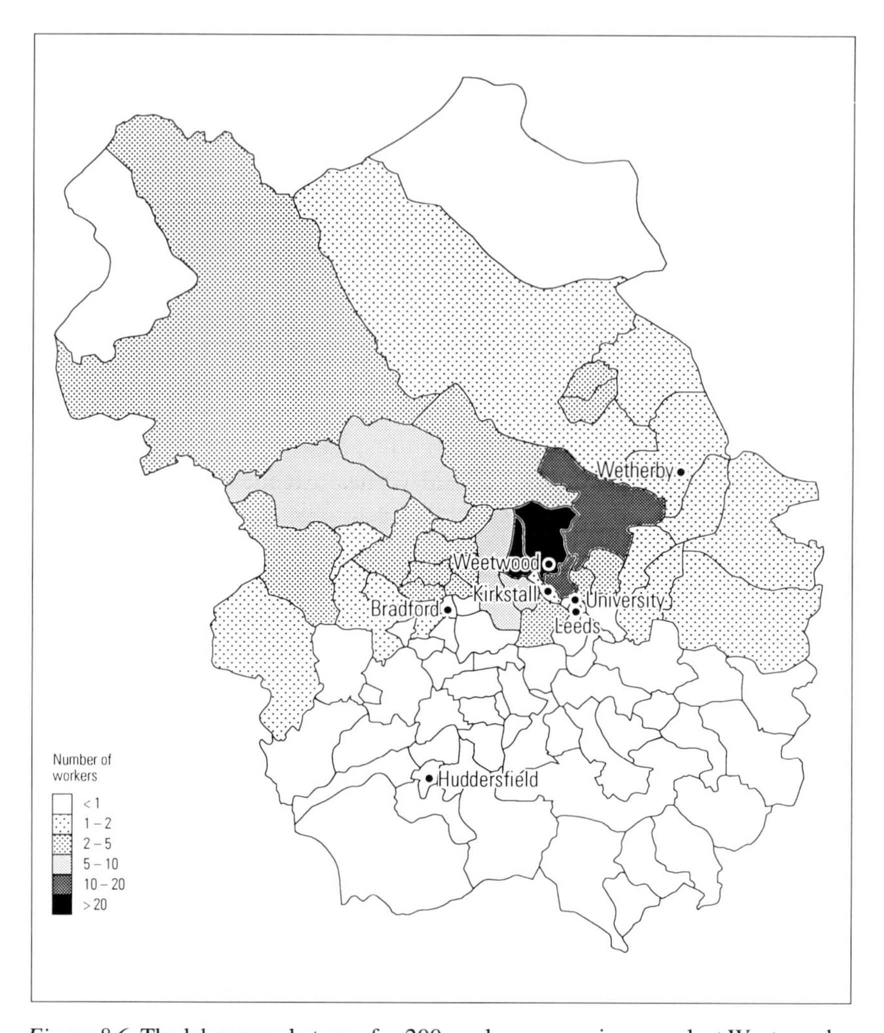

Figure 8.6 The labour market area for 200 workers on a science park at Weetwood

plans for a new science park in Leeds (see Leigh 1991). The research task for the future is much more detailed modelling of the impacts of job losses and gains on the local economy as a whole.

8.5 RETAILING

In this section we describe a performance indicator and modelling approach which has been developed within the School of Geography at the University of Leeds and applied through GMAP Limited, a University-owned company, to a whole spectrum of retail and service activities. We focus in more detail on an application developed between GMAP and WH Smith Limited, one of Britain's leading high street retailers.

8.5.1 Background and context

The spatial interaction modelling approach which has been developed at the University of Leeds over the past twenty years has been well documented elsewhere in this volume (especially Chapters 3 and 7) and in a number of books and papers (e.g. Wilson 1974; Clarke and Wilson 1985a). In the mid-1980s, the University of Leeds began actively to market this modelling approach to potential clients, initially through the University of Leeds Industrial Services (ULIS) Limited, and later through an independent university-owned company, Geographical Modelling and Planning (GMAP) Limited. The first two major clients were Toyota GB and WH Smith Limited, with whom work began in 1987. In both of these cases, new development work is still being undertaken, but GMAP has extended its client base to include many other businesses and public sector clients. These clients range from local authorities, Regional Health Boards and Water Authorities through banks and building societies to leisure activities, supermarket retailing, out-of-town activities and high street retail. Outside the UK, systems have been developed across Europe and in North America and Australia. GMAP has a current workforce of around seventy full-time employees.

In the next section, we describe some of the main features of the system developed between GMAP and one of its earliest clients. A fuller description is provided by Birkin and Foulger (1992).

8.5.2 An application to WH Smith Limited

About the retailer

WH Smith Group PLC was founded in 1792 as a seller of newspapers and books on stations with the advent of the railways. The company has expanded continually over the last two hundred years, and now has interests from news and book wholesaling and office stationery supplies through to high street retailing in both Britain and the United States.

The high street chains include over 400 WH Smith branches whose major product groups are newspapers, books, magazines, stationery, greeting cards, recorded music and videos. The company also owns Our Price Music and Our Price Video, leading specialist music retailers; Waterstones, the market leader in specialist book selling; and Paperchase, a distinctive retailer of high quality stationery and greeting cards.

The aims of the group include being the leading retailer of books, magazines, stationery, recorded music and video in the UK. Most of the markets in which WH Smith are operating are considered to be fairly mature – in other words there is not much potential for growth in the overall size of the market (the obvious but only exception to this being the rapidly expanding video sector). In order to achieve market leadership, WH Smith therefore needs to grow through increasing its market share. These objectives are currently being pursued through heavy investment both in information technology and in expansion of the existing store network into new towns and sites.

It is vital that the programme of new store openings be supported by accurate sales forecasts for those new units for at least three reasons. In the first place, without accurate forecasts it is impossible to plan effectively and produce meaningful budgets or out-turns for the stores. A second reason is that with the average cost of fitting out new stores increasing all the time, and the huge rises in rates in the last few years, it is of paramount importance that resources are not misdirected towards unprofitable new openings. Third, mistakes are difficult to rectify as closures are bad for the image of a high profile retailer such as WH Smith. A similar need for reliable 'what if?' information for planning is the main consideration linking the whole range of clients described in Section 8.5.1 above.

At an early stage, it was recognized that the spatial interaction modelling approach was too complex for the development of a national model. We decided to break Great Britain down into thirty geographical regions. The analysis focuses on the six major product groups offered by WH Smith – news, books, stationery, sounds, greeting cards and video. Expenditure for these different products by postal district is estimated using the Family Expenditure Survey, syndicated market research data and in-house market research data on expenditure patterns for different socio-demographic groups. All of this information is combined with 1991 Census of Population data, updated using techniques developed by GMAP.

Shopping centres within each region are defined as any freestanding retail outlet, or cluster of retail outlets, in which one or more of the main WH Smith products is sold. Over 1,500 centres have been identified in this way. The identification of competing retailers, together with estimates of their size and competitiveness, is achieved using a combination of retail directories, computerized Yellow Pages listings, market intelligence reports and WH Smith in-house data and intelligence.

A rolling programme of implementation was devised, beginning with a pilot study for the West Yorkshire region. Over two years were taken in the model implementation to all thirty regions. The main criterion by which the model was validated was that it should ultimately be capable of consistently forecasting individual product sales by store within ±10 per cent of known revenues, in order to meet the objectives defined above.

The information is presented within a National Intelligence System which combines market data, competitor inventories and store performance information within a geographic information system which allows users to perform database queries, to produce maps and graphs, to perform statistical analyses and, most importantly of course, to undertake 'what if?' model simulations.

Clearly much of the information contained within this National Intelligence System is commercially confidential. For this reason, a demonstration system has been developed in which the data have been changed. In the next section, we will illustrate some of the contents and applications of the National Intelligence System, using this demonstration system.

Benefits of the system

The principal benefit of the system is undoubtedly that it provides a 'what if?' sales forecasting facility. The model enables WH Smith to assess the effect of a new WH Smith Group store anywhere in the country within minutes. The user only has to enter the size of the proposed new store in the chosen town (floorspace is then assigned to product codes according to a standard store layout). The output shows the expected sales for the new store, the effect on stores in surrounding centres (sales deflections) and changes in catchment populations, market size and market share for each product group. The effect of changing the amount of space given to each product within existing branches can also be assessed, as can the effect of new competitor openings.

Nevertheless, there are many further benefits which accrue through the use of performance indicators within a model-based intelligence system. The baseline model generates a market size and catchment population for each of the 1,500 centres identified. For towns where there is a WH Smith Group store, a predicted sales level and market share are generated. All these variables are provided for each of the six major product groups.

These model outputs, together with inputs including competitor presence and floorspace, population profiles, actual WH Smith sales, and many others, are all integrated within the National Intelligence System. A standard menu of indicators is shown in Table 8.12. The National Intelligence System allows the user to interrogate particular towns, or to combine indicators flexibly across a region or for the whole country. It is also possible to perform specialist searches, i.e. to identify all the locations of a particular competitor across the country. There is an enormous range of applications for this quantity and quality of information. Some of the main applications are discussed next.

Market intelligence

The National Intelligence System enables stores to be ranked in order of size, potential catchment, extent of competition, performance (observed versus existing revenue) and so on. Indeed, as we observed above, any of the indicators shown in Table 8.12 can be combined, filtered, ranked and analysed statistically. A typical application is shown in Table 8.13. Here centres in the Avon region have been extracted, with the centre catchment population and floorspace, a social class index, WH Smith floorspace, sales per foot and market share. Note that all floorspace is measured here as effective floorspace

Table 8.12 Performance indicators available in the WH Smith National Intelligence System

Region: Avon

Residential Districts Indicators Selection Screen

Tabulating

Product: All Main

A – Expenditure
B – Expenditure/Head
C – WHS Market penetration
D – OPM Market penetration
E – Wat Market penetration
F – Ppc Market penetration
G – OPV Market penetration
H – Corp Market penetration

Indicators Chosen

I – Pop. change 81–8
J – % Social class
K – Act. Social class
L – % Population
M – Act. Population
N – Geography

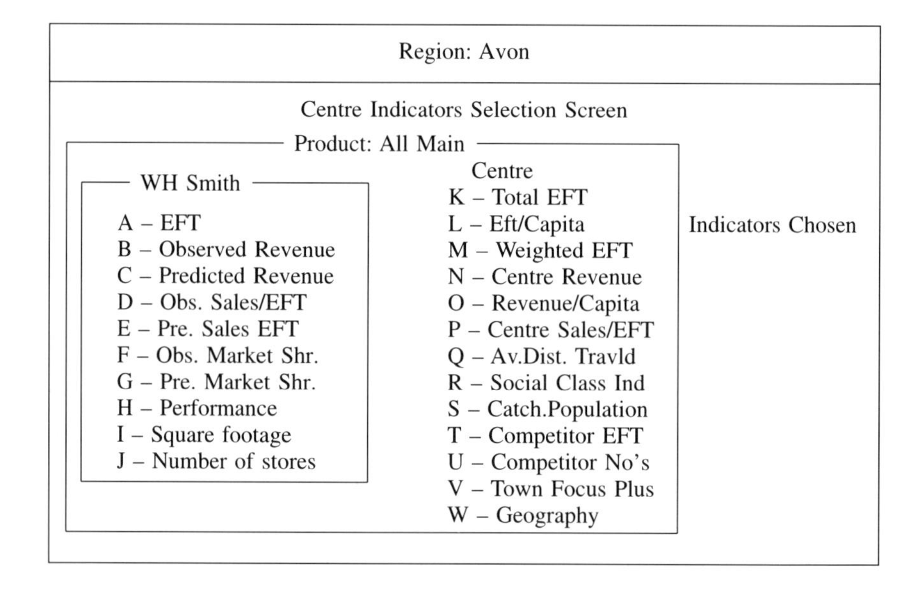

Region: Avon

Centre Indicators Selection Screen

Product: All Main

WH Smith

A – EFT
B – Observed Revenue
C – Predicted Revenue
D – Obs. Sales/EFT
E – Pre. Sales EFT
F – Obs. Market Shr.
G – Pre. Market Shr.
H – Performance
I – Square footage
J – Number of stores

Centre

K – Total EFT
L – Eft/Capita
M – Weighted EFT
N – Centre Revenue
O – Revenue/Capita
P – Centre Sales/EFT
Q – Av.Dist. Travld
R – Social Class Ind
S – Catch.Population
T – Competitor EFT
U – Competitor No's
V – Town Focus Plus
W – Geography

Indicators Chosen

total, which is a measure of the shelf space allocated to products. Also note that the social class index is simply the percentage of the catchment population who are in social class A or B. The data in Tables 8.13, 8.14 and 8.15 have been randomized to protect client confidentiality.

The table is ranked by centre size, showing a Smith's presence in the nine largest centres, and also Wells, Clevedon and Wellington. In general, WH Smith sales per foot is higher in the larger centres but the market share is lower. The centre which has the highest catchment population but lacks a WH Smith store is Bedminster. The effect of opening a new store in Bedminster

Table 8.13 Key centre performance indicators for the Avon region

				WH Smith, all main products		
Name	*All main catchment population*	*All main total EFT*	*Social class*	*EFT*	*Sales/EFT (obs.)*	*Market share (obs.)*
Bath	279,477	12,050	36.6	590	3,610.7	8.0
Bristol	372,619	9,498	33.2	790	4,356.8	12.5
Taunton	118,231	5,805	35.0	849	2,966.8	24.0
Clifton	84,273	5,745	36.5	588	2,410.5	13.6
Yeovil	98,416	4,389	33.7	535	3,327.3	21.8
Weston Super Mare	78,152	3,800	34.6	455	2,477.6	19.2
Bridgwater	57,351	3,196	32.2	329	1,510.9	16.1
Kingswood	60,415	2,823	32.5	381	1,291.1	21.5
Trowbridge	50,942	2,204	34.1	369	1,184.3	17.0
Keynsham	19,305	2,168	33.4	0	0.0	0.0
Frome	9,045	1,915	32.8	0	0.0	0.0
Bedminster	29,911	1,814	31.5	0	0.0	0.0
Brislington	11,826	1,779	30.5	0	0.0	0.0
Wells	27,131	1,635	38.7	259	1,071.0	18.1
Bridport	14,823	1,614	39.6	0	0.0	0.0
Fishponds	22,757	1,495	33.3	0	0.0	0.0
Bishopston	19,139	1,350	36.2	0	0.0	0.0
Thornbury	3,694	1,324	37.0	0	0.0	0.0
Burnham-on-Sea	12,064	1,222	36.1	0	0.0	0.0
Melksham	10,346	1,181	33.6	0	0.0	0.0
Yate	9,774	1,175	36.1	0	0.0	0.0
Clevedon	15,897	1,161	44.5	149	1,161.7	12.0
Shirehampton	6,147	1,109	32.9	0	0.0	0.0
Nailsea	6,677	1,105	43.7	0	0.0	0.0
Westbury Park	10,884	1,039	37.8	0	0.0	0.0
Sherborne	9,603	1,035	38.6	0	0.0	0.0
Henleaze	9,628	983	36.4	0	0.0	0.0
Glastonbury	8,330	975	34.8	0	0.0	0.0
Crewkerne	12,742	939	34.5	0	0.0	0.0
Wellington	10,511	937	36.4	181	534.8	16.9
Chard	11,046	923	34.9	0	0.0	0.0
Eastville	7,336	921	32.7	0	0.0	0.0
Lyme Regis	7,521	868	41.5	0	0.0	0.0
Midsomer Norton	2,929	830	36.7	0	0.0	0.0
Henbury	3,746	696	35.4	0	0.0	0.0
Staple Hill	5,106	663	33.3	0	0.0	0.0
Worle	3,752	661	35.8	0	0.0	0.0
Bishopsworth	3,608	547	29.1	0	0.0	0.0
Wincanton	3,027	547	33.6	0	0.0	0.0
Portishead	2,700	512	43.0	0	0.0	0.0
Street	11,298	508	33.7	0	0.0	0.0
Ilminster	660	465	36.2	0	0.0	0.0
Bradford-on-Avon	836	405	36.0	0	0.0	0.0
Williton	36	398	38.0	0	0.0	0.0

Table 8.13 continued

Name	*All main catchment population*	*All main total EFT*	*Social class*	*WH Smith, all main products* *EFT*	*Sales/EFT (obs.)*	*Market share (obs.)*
Cheddar	476	395	43.6	0	0.0	0.0
Redfield	3,866	389	31.0	0	0.0	0.0
Horfield	2,680	384	35.0	0	0.0	0.0
Chipping Sodbury	1,019	357	36.3	0	0.0	0.0
Westbury	1,747	346	31.7	0	0.0	0.0
Somerton	1,572	335	37.4	0	0.0	0.0
Shepton Mallet	3,123	331	36.2	0	0.0	0.0
Stapleton	1,051	305	33.2	0	0.0	0.0

Note: EFT, effective floorspace total.

Table 8.14 Revenue forecast for a new store at Bedminster

Product: All main	*Current centre: Bedminster* *WHS*	*OPM*	*Wat*	*PPC*	*OPV*	*CORP*	*Centre*
Number of stores	1	0	0	0	0	1	
EFT (baseline)	0	0	0	0	0	0	1,814
EFT (model)	345	0	0	0	0	345	2,159
Sales (£,000, baseline)	0.0	0.0	0.0	0.0	0.0	0.0	1,505.6
Sales (£,000, model)	448.5	0.0	0.0	0.0	0.0	448.5	1,794.1
Sales/EFT (£, baseline)	0.0	0.0	0.0	0.0	0.0	0.0	830.0
Sales/EFT (£, model)	1,300.1	0.0	0.0	0.0	0.0	1,300.1	831.0
Market share (baseline)	0.0	0.0	0.0	0.0	0.0	0.0	
Market share (model)	25.0	0.0	0.0	0.0	0.0	25.0	
Catchment (baseline)	29,911	Average distance (km, baseline) 9.27					
Catchment (model)	38,298	Average distance (km, model) 9.32					

Note: WHS, WH Smith; OPM, Our Price Music; Wat, Waterstones; PPC, Paperchase; OPV, Our Price Video, Corp, Group.

Table 8.15 Deflections from Bristol and Clifton stores

	News	*Books*	*Stationery*	*Sounds*	*Cards*	*Video*	*All*
WH Smith (total, £,000)	–5.8	–4.4	–6.8	–12.3	–1.5	–12.6	–43.4
WH Smith (%)	–0.1	–0.2	–0.5	–1.3	0.0	–0.3	–0.3
Our Price Music (total, £,000)	0.0	0.0	0.0	–5.4	0.0	–10.5	–15.9
Our Price Music (%)	0.0	0.0	0.0	–1.4	0.0	–0.2	–0.3
Waterstones (total, £,000)	0.0	–23.3	0.0	0.0	0.0	0.0	–23.3
Waterstones (%)	0.0	–0.4	0.0	0.0	0.0	0.0	–0.4
Paperchase (total, £,000)	0.0	0.0	–3.3	0.0	–0.1	0.0	–3.4
Paperchase (%)	0.0	0.0	–1.1	0.0	0.0	0.0	–0.6
Our Price Video (total, £,000)	0.0	0.0	0.0	0.0	0.0	–0.1	–0.1
Our Price Video (%)	0.0	0.0	0.0	0.0	0.0	–0.1	–0.1
Total difference (£,000)	–5.8	–27.7	–10.1	–17.7	–1.6	–23.2	–86.1
Total (%)	–0.1	–0.4	–0.6	–1.3	0.0	–0.3	–0.3

is shown in Tables 8.14 and 8.15. Table 8.14 shows a revenue forecast for a unit of 3,000 square feet of £448,500 per annum. Table 8.15 shows that £86,000 of this revenue is deflected from existing WH Smith stores (mainly Bristol and Clifton). Figure 8.7 shows market penetrations in the Avon region before and after the new store opening is simulated. The right-hand map shows the existing penetrations across the whole region. The left-hand map shows an expanded window around Bristol, with a new store in Bedminster. The figure illustrates that the new store helps to raise market penetrations in the Bristol area to the levels found already around Bath, Yeovil and Taunton.

There are many other types of application. For example, in many large towns, WH Smith already has more than one store. The system can identify the top ten towns where there is potential for a second WH Smith store – towns where the level of competition per head of catchment population is lower than average, WH Smith sales per unit of floorspace are relatively high and WH Smith's share of the market for each product is relatively low.

Stores vulnerable to competitive entry can also be identified (perhaps where the WH Smith market share is abnormally high in certain product areas). From this kind of analysis, a 'what if' scenario can be run to test the effect on the WH Smith business of a competitor opening, and this may spark a search for a site for one of Smith's own competitive chains.

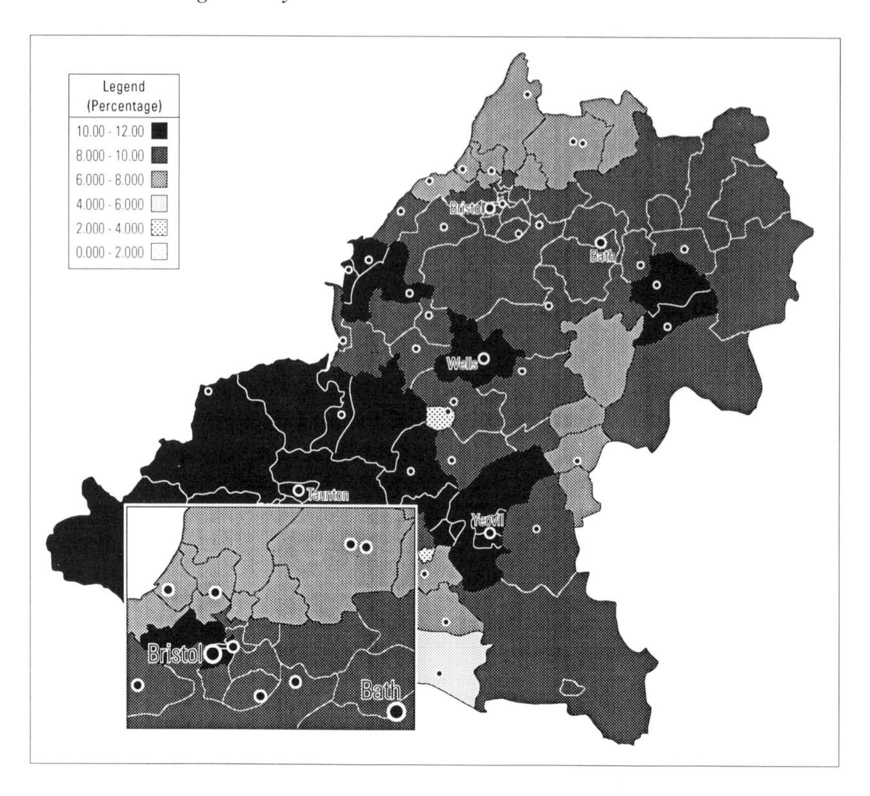

Figure 8.7 Market penetrations before and after (inset) the new store at Bedminster, near Bristol (○)

Current store performance

The National Intelligence System can be used to look at the difference between the model prediction of WH Smith sales in an existing store and the actual sales. Modelling error has been established to be less than 10 per cent; hence branches where the predicted sales are more than 10 per cent away from the real figures can be examined to isolate the reasons for the under- or over-performance. Such analysis may lead to a recommendation that the branch be re-sited or extended or, in rare cases, closed.

Demographic analysis

The intelligence database shows the number and percentage of people in any postal district who are in any age, sex or social class band, together with the average expenditure per head on WH Smith products of those residents. This information has great potential as a local promotions

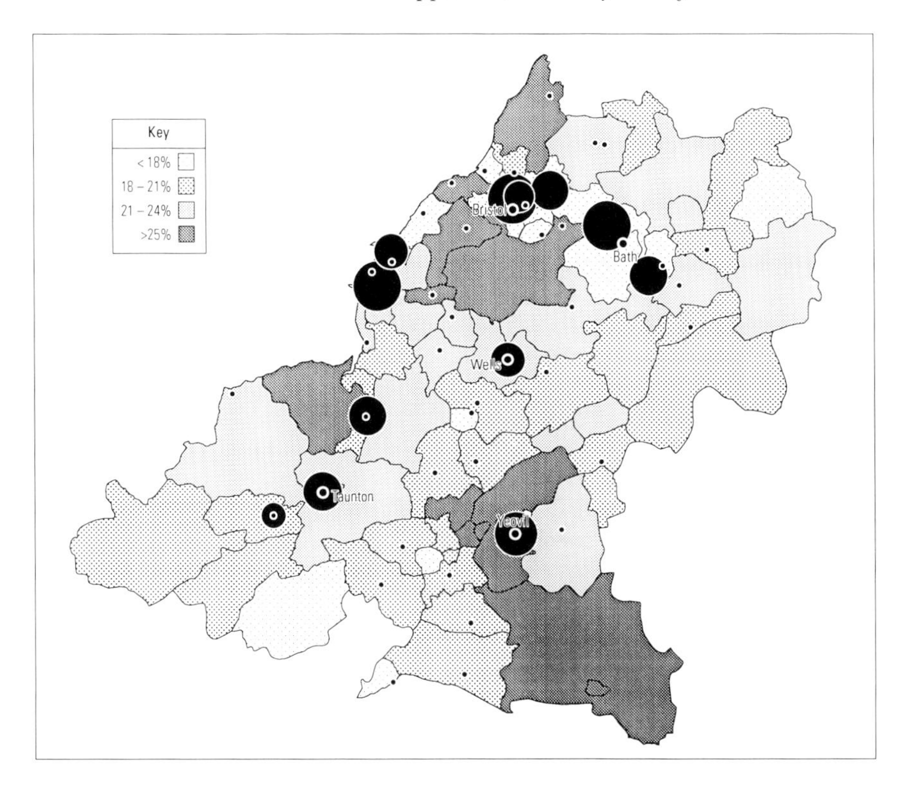

Figure 8.8 Store centre size overlaid with percentage of population in social group A

tool when certain sectors of the population need to be targeted. In the latest version of the system it is now possible to overlay demographic data relating to postal districts with data for centres or outlets. An example of this is shown in Figure 8.8, where centre sales of books per foot is overlaid on the proportion of the population in social class AB. Were the data not altered for the protection of confidentiality (see previous section) one might expect to see a fairly strong relationship here.

Summary

As we observed above, the ability to provide accurate sales forecasts for new stores is of fundamental importance to retail businesses. In addition, however, the specification and use of appropriate performance indicators can provide further benefits of great tactical and strategic significance. The indicators which are useful in this regard include both outlet-based indicators (such as store performance, market shares within a centre and catchment-based indicators) and residence-based indicators (such as market penetration and provision ratios).

GMAP's National Intelligence System provides WH Smith with a proven and reliable mechanism for sales forecasting. In addition, the system provides the company with the most comprehensive set of market intelligence data it has ever had. Although the investment made by the company has been costly in terms of both time and money, it is recognized that this investment can be fully recouped by siting just two stores profitably where they might previously have been disasters.

9 An application of performance indicators in Italy

S. Occelli

9.1 INTRODUCTION

The theoretical and methodological advances in the conceptual development of performance indicators has led quite naturally to a desire to put them to practical applications in order to test their real effectiveness as analytical tools. In this chapter we describe a recent application commissioned in the early 1990s by one of the Italian regional planning authorities for use as an analytical tool. The package (which we refer to here as the RPA package) was conceived as a flexible planning tool for the organization and interpretation of spatially defined socio-economic information. Both the broader specification and the experience acquired in the development of a previous package (Bertuglia *et al.* 1989a, 1991a; Gallino *et al.* 1990) suggested the opportunity to abandon a form linked exclusively to specific aspects of the functional and structural organization of an urban system. It was thus decided to opt for an approach which favoured flexibility of application (in terms of databases, categorization of variables and zoning). This led to what could be defined as an 'open package' in which the indicator is combined or assimilated with a mathematical 'operator' which processes the data on the basis of a defined set of rules. The exact implications of this will become clearer in the examples given in Section 9.3.

First, we describe the main features of the package, looking at the general architecture and operational characteristics, without entering into the details of single indicators. We then present some results of the application of the package to the region of Piedmont, carried out as part of the testing phase. Finally, on the basis of this experimentation, we make some general comments about future prospects in relation to three basic aspects – the conceptual design of the package, its practical features and the computer technology adopted.

9.2 THE INDICATOR PACKAGE

In constructing the package it was assumed that the subject of the investigation would be a geographically defined system possessing a given structure of functional interdependences and spatial interrelations between activities (e.g. between place of residence and place of work or between the population and a particular service). We are interested in the spatial configuration of the system expressed in terms of the distribution of the various activities in relation to the zones and the performance of the system in relation to its functional and spatial structure.

The indicators were developed from a number of standard indicators based on the classification introduced in Chapter 4 – namely, descriptive, profile and performance indicators. The set of indicators proposed is by no means exhaustive but is intended to serve as an initial nucleus to be extended and improved at a later date. The purpose of the indicators was primarily both to provide a descriptive profile of a series of phenomena in a given geographical context and to help in the definition of problem areas.

The information system can be divided into two different kinds of module. These are:

1 analysis modules;
2 calculation modules.

The former relate to the subjects (systems) being analysed. The latter concern the different kinds of indicator used and in particular the operations involved in their calculation. The division into two distinct kinds of module was a response to the need to provide an overall picture of the range of analytical operations which could be carried out, or the different kinds of indicator which could be applied, and the need to guarantee a high degree of flexibility and interactivity in the management of the basic information without creating an excessive computational burden (in terms of time, file management etc.).

The principal feature of this package is the function of the indicator, conceived of as a 'calculation operator' which acts on a given set of information selected by the user according to his or her requirements (subject of course to certain predefined rules). Although this means that the package does not provide predefined sets of indicators for each subsystem, there are enormous advantages in terms of management, updating and organization of the database. It also provides far greater operational flexibility, which is particularly important if it is to be used with personal computers.

A further extremely significant feature of this package is that it requires much greater involvement on the part of the user. It is he or she and not the package which decides which aspect of the phenomena is important to investigate, what kind of indicator to apply and which information to utilize. The structure of the package is shown diagrammatically in Figure 9.1. The main module has a central role, containing the set of basic indicators which are applicable for the private, public and general urban system modules. The urban subsystem modules contain a further (though in this version limited) series of indicators more specific to the problems of these systems.

As Figure 9.1 shows, the indicator modules are based on the ideas raised in Chapter 4 – namely the three indicator modules of description, profile and performance.

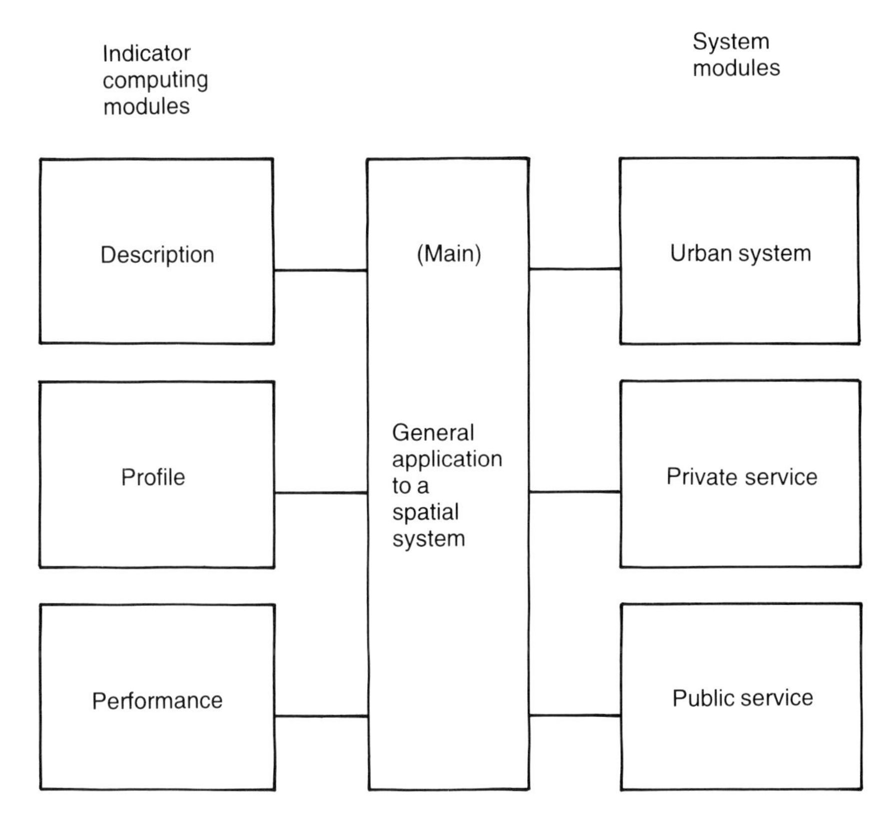

Figure 9.1 Diagram of the package structure

The first module is for the calculation of *descriptive* indicators. This contains utility indicators (involving elementary operations on the basic data), comparative indicators (performing operations designed to highlight zonal differences) and graphic representation indicators (making it possible to map the spatial distribution of any indicator for any zoning desired). These indicators can be applied either to raw data or to the output of the calculation of the other indicators. They can also be applied to all the various analysis modules.

The second module is for the calculation of descriptive *profile* indicators. These perform calculations aimed at identifying typical features of an area from the point of view of the interrelations, degrees of similarity and levels of specialization.

The third module is for the calculation of *performance* indicators. These involve calculations designed to provide a measure of the 'performance' associated with a particular spatial distribution of a given activity or service. For these indicators a subdivision was made according to type (those

Table 9.1 Indicator computing modules and main category indicators

Computing modules	*Category*	*Description of the main indicators*
Description	Utility indicators	Fundamental algebraic operations (sum, product, ratio etc.)
	Comparative indicators	Row and column percentages, row mean and modal values, column minimum and maximum values, column variances and normalizations
	Mapping indicators	Automatic classifications (sixteen classes) Automatic classifications into three classes according to the distribution around the mean value Manual classification (chosen by the user)
Profile	Relationships indicators	Flow generation ratios Self-contained ratios Mobility structure indices
	Homogeneity indicators	Simple index (referring to a single activity distribution: ratio of the zonal variance to the total variance) Composite index (referring to a set of activities: median value of the homogeneity simple index values)
	Specialization indicators	Zonal index (ratio of the zonal percentage activity to the average) Relative zonal index (ratio of the zonal percentage to the mean) Selection index (showing the activity whose weight is significantly higher than the mean value)
Performance	Accessibility indicators	Hansen-type accessibility indices (according to the form of the distance function, transportation mode, way of perceiving the distance friction, zonal autocontribution)
	Influence indicators	Field opportunity indices (specified in the same way as the accessibility indicators)
	Independence indicators	Indices obtained as relative ratio of influence and accessibility indicators
	Potential delivery and catchment population indicators	(see Chapter 5)

Efficiency and effectiveness indicators	(see Chapter 5)
Transport cost indicators	Generalized, mean and per capita mean travel cost indicators (articulated by origin and destination, and by transport mode) Zonal autocontribution indices Weighted mean generalized transport cost

measuring accessibility, influence, independence, potential level, transport costs and so on). The content of these modules is outlined in Table 9.1.

A further specific feature of the package concerns the treatment of the spatial dimension, i.e. the zoning of the study area. One of the most common problems of geographical analysis concerns the need to operate at different scales. This more often than not necessitates laborious operations of aggregation and reorganization of data in order to arrive at the desired level of spatial resolution. In this respect the present package offers two advantages: first, the automatic reaggregation of information to any desired scale (from the finest zoning division upwards), and second, the possibility of selecting any given set of zones from the basic zoning system.

9.3 AN APPLICATION OF THE INDICATOR PACKAGE

We now illustrate and comment on an application of the RPA indicator package to Piedmont. The application was carried out primarily in order to test the overall design (i.e. to determine whether the set of indicators developed was sufficiently comprehensive to permit an exploratory analysis of the region) and the robustness and viability of the package (the adequacy of the software).

For the purposes of this study the region was divided into a relatively small number of zones (nineteen in all) which were a subdivision of six main administrative provinces (Figure 9.2) (provincial capitals are shown in capital letters). The provinces themselves represented a more aggregate level of zoning. The package can in fact operate with several hundred zones, up to a theoretical maximum of 2,000. The upper limit depends on the capacity of the computer used.

The choice of this rather large-scale zoning system was made for three reasons: first, the fact that a certain number of databases were already available for these zones; second, to establish to what extent the package would permit us to obtain an 'information gain' in relation to other more traditional methods of analysis; and finally, to see whether it would permit us to express the various facets of the system we were interested in exploring.

We present here a selection of examples from each of the categories of

Figure 9.2 Map used for digitization showing subareas and provinces

indicators shown in Table 9.1. They are presented in graphical or tabular form with the relevant details summarized below. These include the category and name of the indicator, its mathematical expression (where relevant), the type of information considered and a brief comment on the results.

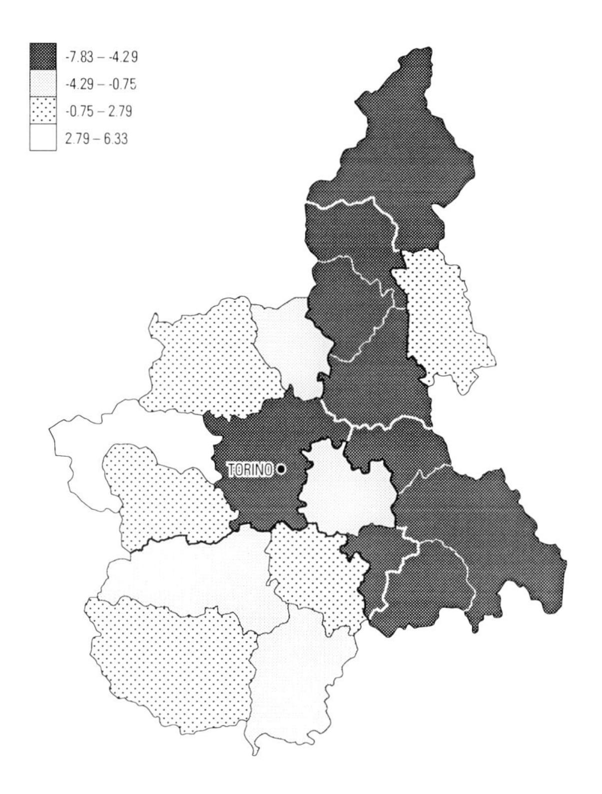

Figure 9.3 Percentage variations in population 1981–91

Category: comparative indicator (description module)
Indicator: percentage variation
Information: population distribution in 1981 and 1991 (*Source*: Population Census)
Remarks: This is a relatively straightforward indicator commonly applied to stock variables. In this case it highlights the population decline which spreads out from the metropolitan core along a north–south axis in the eastern part of the region. The rest of the region shows relative stability or a slight increase in population over this period. The classes are computed automatically.

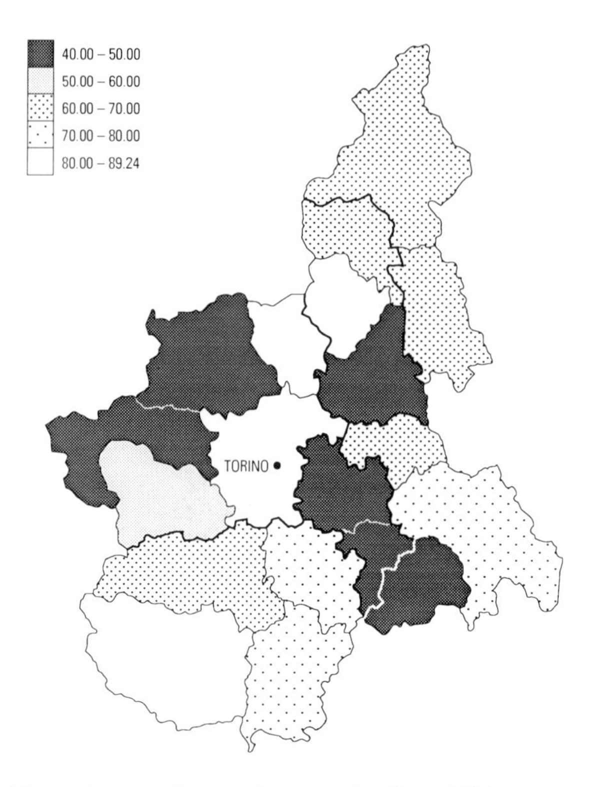

Figure 9.4 Self-containment of outward commuting flows 1989

Category: relationship indicator (profile module)
Indicator: self-containment

$$U_i = \frac{T_{ii}}{\Sigma_j T_{ij}} \times 100 \qquad (9.1)$$

where T_{ij} are the commuting flows from zone i to zone j

Information: estimates of rush-hour traffic flows for 1989 obtained on the basis of 1981 census data

Remarks: The results show that seven of the nineteen zones have a self-containment index below 50 per cent. This indicates, from the point of view of residence, a relatively high level of interaction with the other zones. These represent the less densely populated areas of the region, however, situated in close proximity to zones offering greater opportunity in terms of jobs and services.

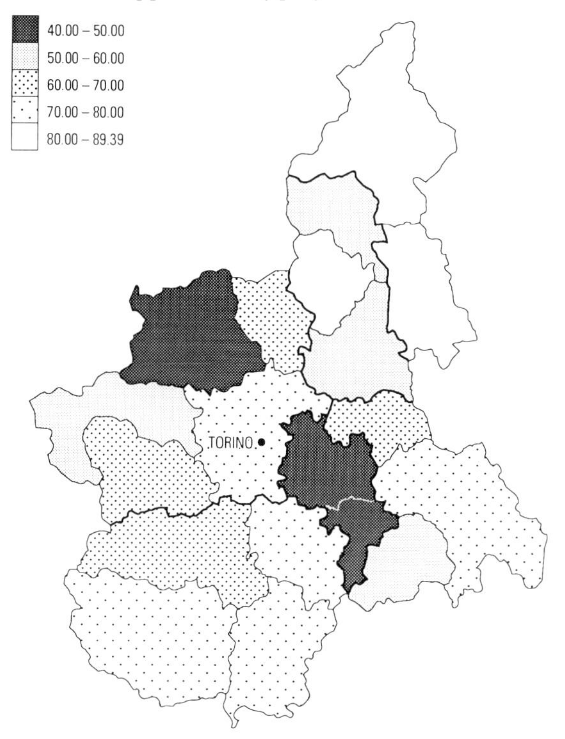

Figure 9.5 Self-containment of inward commuting flows 1989

Category: relationship indicator (profile module)
Indicator: self-containment

$$E_j = \frac{T_{jj}}{\Sigma_i T_{ij}} \times 100 \qquad (9.2)$$

where T_{ij} are commuting flows from zone i to zone j.

Information: as for (9.1)

Remarks: The pattern identified in Figure 9.4 remains, with few exceptions, basically unvaried. In the case of the inward flows the values are slightly higher, indicating that in relation to job supply (place of work) the subareas are relatively self-contained. The crucial question for indicators of this type is the threshold adopted for the definition of 'self-containment' (usually 75 per cent). This decision will depend on the objectives of the analysis and the system being studied. It should be noted that the concept of self-containment, and hence this kind of indicator, is extremely useful for the identification of spatial subsystems in transport analysis and the definition of employment catchment areas for labour market analyses (see also Chapter 8).

Table 9.2 Zonal specialization index by activity sector 1991

	Retail	*Industry*	*Private sector services*	*Public sector services*
Novara	1.0336	1.0143	0.8857	1.0592
Vercelli	0.9497	1.1392	0.8233	0.8925
Torino	0.9528	1.0104	1.0643	0.9516
Alessandria	1.1047	0.8967	1.0182	1.1331
Asti	1.0872	0.9321	1.0510	1.0138
Cuneo	1.1167	0.9352	0.9104	1.1436

Category: specialization indicator (profile module)
Indicator: zonal specialization index

$$S_i^m = \frac{A_i^m \;/\; \Sigma_m \, A_i^m}{\Sigma_i \, A_i^m \;/\; \Sigma_m \, \Sigma_i \, A_i^m} \tag{9.3}$$

where A_i^m is the number of jobs in sector m in zone i

Information: jobs by sector in 1991: retail, industry, private sector services, public sector services (Trades and Industry Census)

Remarks: This indicator is an index of the spatial concentration of an activity, measured in relation to its total level in the system. In general the higher the index is, the more we can consider the zone to be 'specialized'.

Table 9.3 Relative specialization index by activity sector 1991

	Retail	*Industry*	*Private sector services*	*Public sector services*
Novara	0.9931	1.0266	0.9237	1.0260
Vercelli	0.9125	1.1530	0.8587	0.8646
Torino	0.9154	1.0227	1.1100	0.9218
Alessandria	1.0615	0.9076	1.0619	1.0977
Asti	1.0446	0.9434	1.0962	0.9821
Cuneo	1.0729	0.9466	0.9495	1.1078

Category: specialization indicator (profile module)
Indicator: relative zonal specialization index

$$\bar{S}_i^m = \frac{A_i^m / \Sigma_m A_i^m}{\Sigma_i \left(A_i^m / \Sigma_m A_i^m \right) / I} \tag{9.4}$$

where the variables are the same as those defined in Table 9.2 and I is the total number of zones

Information: as for (9.3)

Remarks: The difference between this indicator and the previous one is that, whereas in Table 9.2 the reference (denominator) was the sectorial composition in relation to the whole region, here it is the mean value of the zonal distribution. The latter is therefore more sensitive to zonal differences in the spatial distribution of the activity.

Even though in reality the difference between the results is slight, we observe that here the values obtained are lower for the retail and public service sectors and higher for the other two, thus emphasizing their spatial concentration. In this sense the indicator shows itself to be more selective in reflecting the distribution characteristics of activities.

Table 9.4 Accessibility of population 1991

	Net accessibility (1)	*Simple accessibility* (2)	(1)/(2)
Verbania	18,083.5742	25,559.2402	0.7075
Novara	167,859.3438	233,055.4375	0.7203
Borgosesia	152,211.2656	158,992.1094	0.9574
Biella	152,105.9688	183,117.8750	0.8306
Vercelli	336,181.7500	363,282.8125	0.9254
Ivrea	298,783.0000	330,342.4063	0.9045
Ciriè	487,332.6250	501,043.6563	0.9726

Category: accessibility (performance module)

Indicator: Hansen-type accessibility index (net and simple)

net accessibility $$A_i = \sum_j P_j \exp(-\beta c_{ij}) \qquad i \neq j \qquad (9.5)$$

simple accessibility $$\bar{A}_i = \sum_j P_j \exp(-\beta c_{ij}) \qquad (9.6)$$

where P_j is the population of zone j, c_{ij} is the travel cost from zone i to zone j and β is the distance impedance parameter

Information: population distribution in 1991 (Population Census); matrix of travel times (for private transport) between subareas

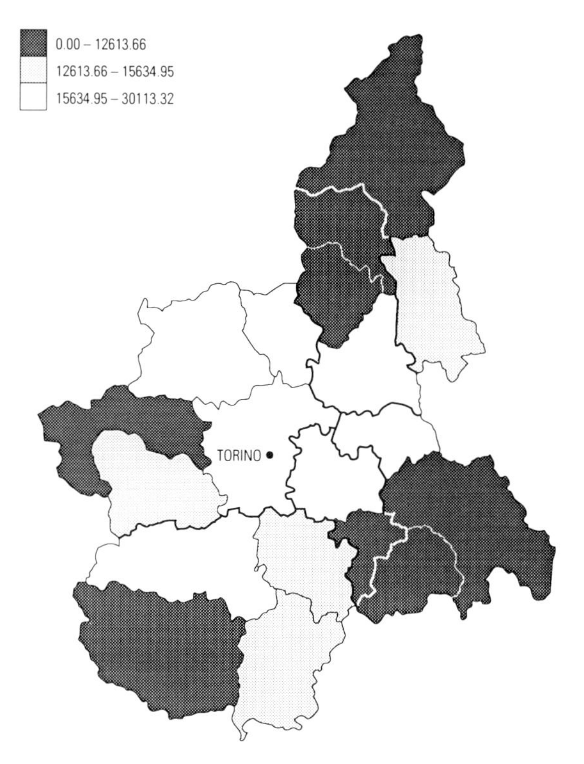

Figure 9.6 Accessibility for public sector service employment 1991

Category: accessibility (performance module)
Indicator: opportunity index

$$A_i = \sum_j E_j \exp(-\beta c_{ij}) \tag{9.7}$$

where E_j are jobs in the public services sector in 1991, c_{ij} is the travel time (for private transport) from zone i to zone j and β is a distance impedance parameter
Information: jobs in public sector services (Trades and Industry Census)
Remarks: The three classes are defined according to their position relative to the mean value of the zonal distributions (below, around or above the mean).

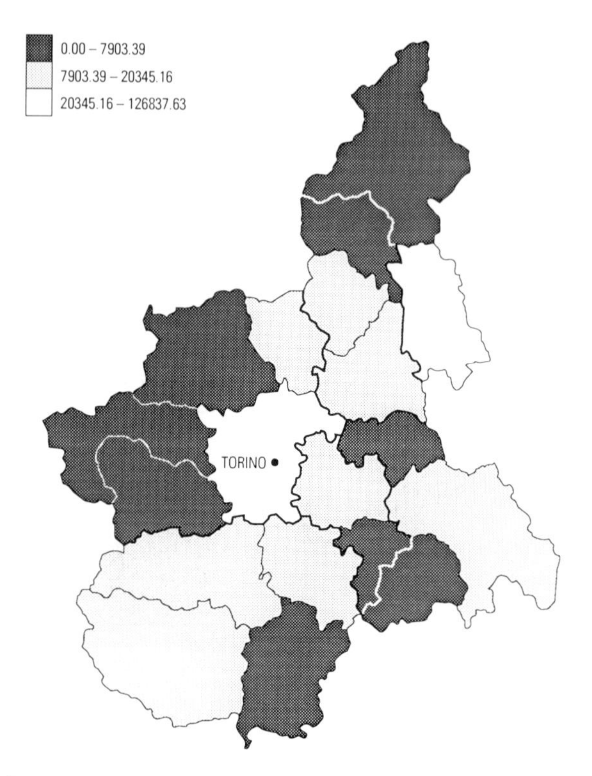

Figure 9.7 Influence indicator for public sector service employment 1991

Category: influence (performance module)
Indicator: field opportunity index (simple)

$$I_j = \sum_i E_j \exp(-\beta c_{ij}) \tag{9.8}$$

where E_j are jobs in the public service sector in 1991, c_{ij} is the travel time from zone i to zone j and β is a distance impedance parameter

Information: see comments on the accessibility indicator (Figure 9.6)

Remarks: This indicator is complementary to the accessibility indicator. It expresses the 'field of influence' of a zone on the surrounding zones in terms of the degree of concentration of an activity and the structure of the spatial interrelations with other zones.

In this case it highlights the importance of two zones – the one in which the regional 'capital' Turin is located and a peripheral zone on the borders of the Lombardy region.

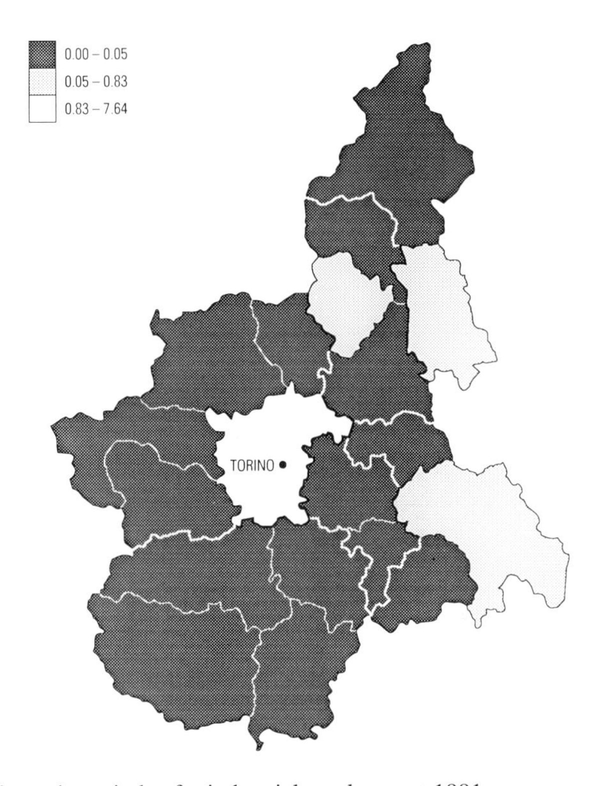

Figure 9.8 Independence index for industrial employment 1991

Category: independence (performance module)
Indicator: influence/accessibility ratio

$$\mathrm{IN}_i^m = \frac{A_i^m \exp(-\beta c_{ii}) \; \Sigma_j \; A_i^m \exp(-\beta c_{ij})}{\Sigma_j \; A_j^m \exp(-\beta c_{ji}) \; \Sigma_j \; A_j^m \exp(-\beta c_{jj})} \quad j \neq i \tag{9.9}$$

where A_j^m are jobs in sector m, c_{ij} is the travel cost from zone i to zone j and β is a distance impedance parameter

Information: see comments on the accessibility indicator (Figure 9.6)

Remarks: The aim of this indicator is to identify, for a specific activity, the role played by a given zone through the combined effect of the influence and accessibility, taking into account the competition with the other zones of the system.

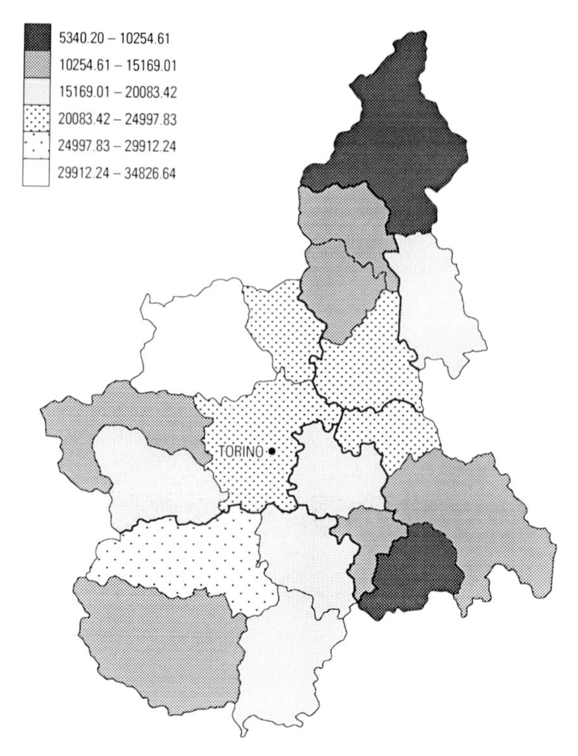

Figure 9.9 Potential delivery for retail activities 1991

Category: potential delivery (performance module)
Indicator: (see Chapter 5)

$$D_i = \frac{\Sigma_j \; F_{ij} \; R_j}{\Sigma_i \; F_{ij}} \tag{9.10}$$

where R_j are jobs in the retail sector in zone j in 1991 and F_{ij} is the number of rush hour trips from zone i to zone j

Information: distribution of jobs in retail activities 1991 (Industrial Census); trip matrix

Remarks: For the construction and valid interpretation of this indicator the soundness of the basic information is particularly important. Although, at least for a regional system, the employment data in the retail sector can be considered an acceptable proxy for the supply of retail services, commuting flows (or the structure of such a matrix) cannot in any way be considered to approximate the pattern of shopping trips, and therefore specific data relating to these flows is necessary (see Chapter 8 for sounder indicators on retail flows).

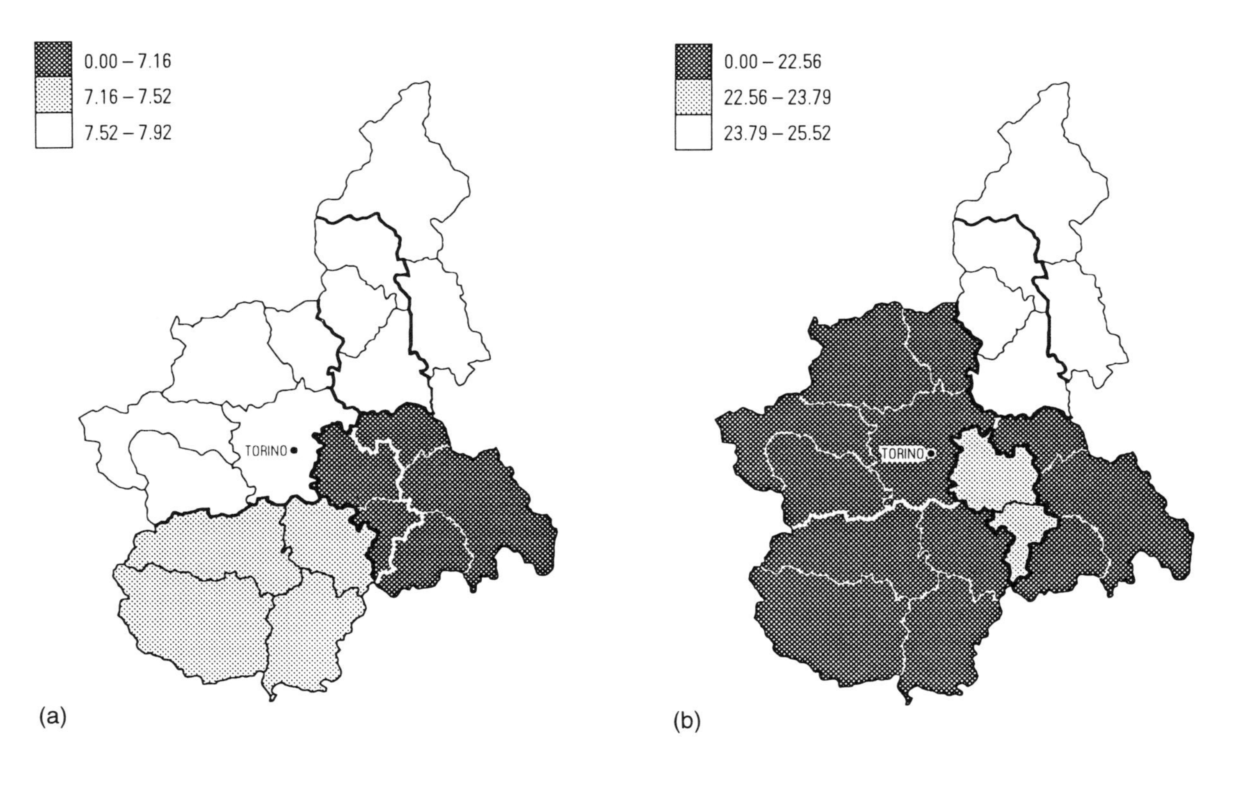

Figure 9.10 Value added per inhabitant (based on place of production): (a) 1980; (b) 1989

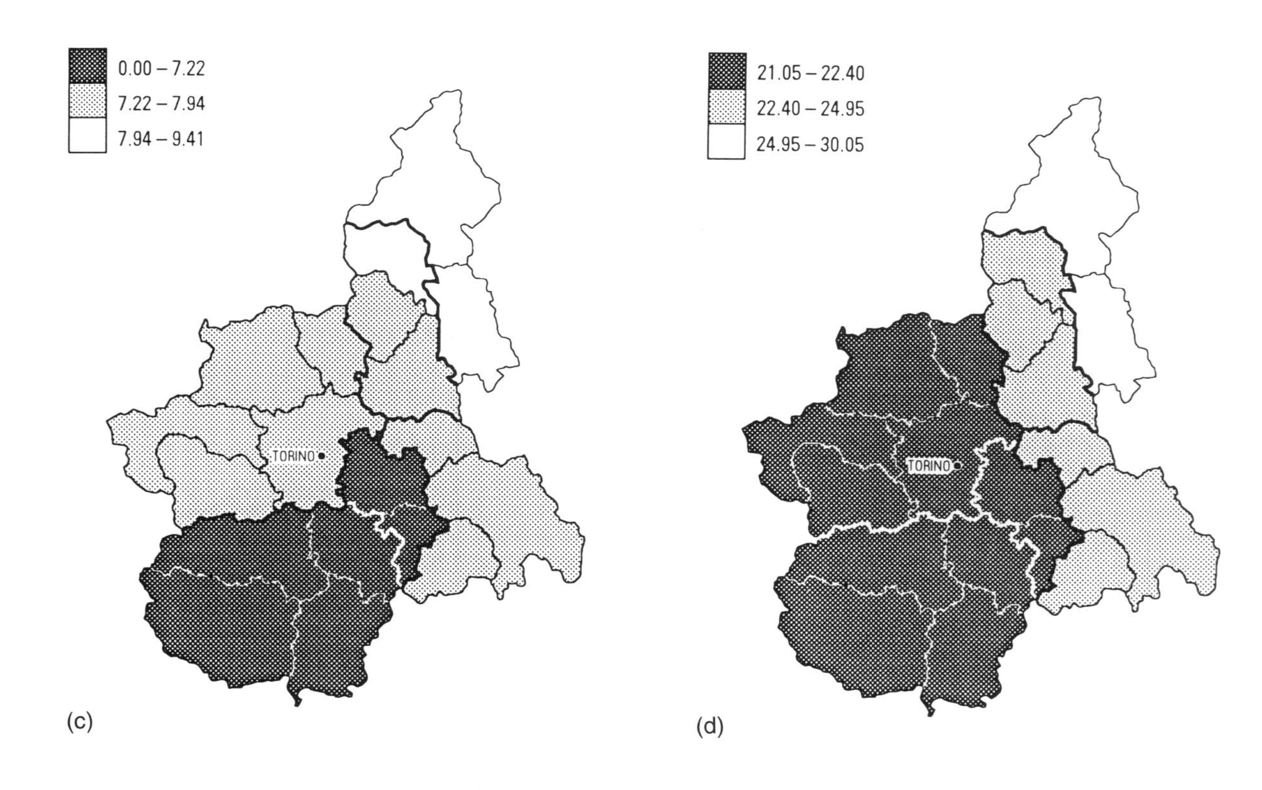

Figure 9.10 (cont) Value added per inhabitant (based on place of residence): (c) 1980; (d) 1989

The value added per inhabitant is a commonly used indicator of the wealth produced by an economic system. Traditionally the indicator is calculated using the ratio

$$W_j = \frac{W_j}{P_j} \tag{9.11}$$

where W_j is the value added for a product in zone j and P_j is the total population in zone j. It assumes implicitly that the wealth produced in a given zone is completely consumed/distributed in that same zone, i.e. the zone is considered to be a completely closed system. Even though this hypothesis may be reasonable on the larger scale (assuming that the larger the area considered is, the greater is the degree of self-containment), it is less likely to be valid when there is a finer level of resolution.

Category: effectiveness (performance module)
Indicator: (see Chapter 5)

$$e_i = \left(\frac{\Sigma_j \, F_{ij} W_j}{\Sigma_i \, F_{ij}} \right) \bigg/ P_i \tag{9.12}$$

where W_j is the value added produced in zone j, F_{ij} are commuting flows from zone i to zone j and P_i is the total population resident in zone i

Information: as for (9.11)

Remarks: Figure 9.10 illustrates the differences which emerge in the results depending upon whether value added is expressed with reference to the place of production or consumption (residence).

In the examples given here, where the basic unit is the province (i.e. a relatively coarse scale), although the differences in the values themselves are not very great the contrast in wealth ranking between the provinces nevertheless shows up quite clearly.

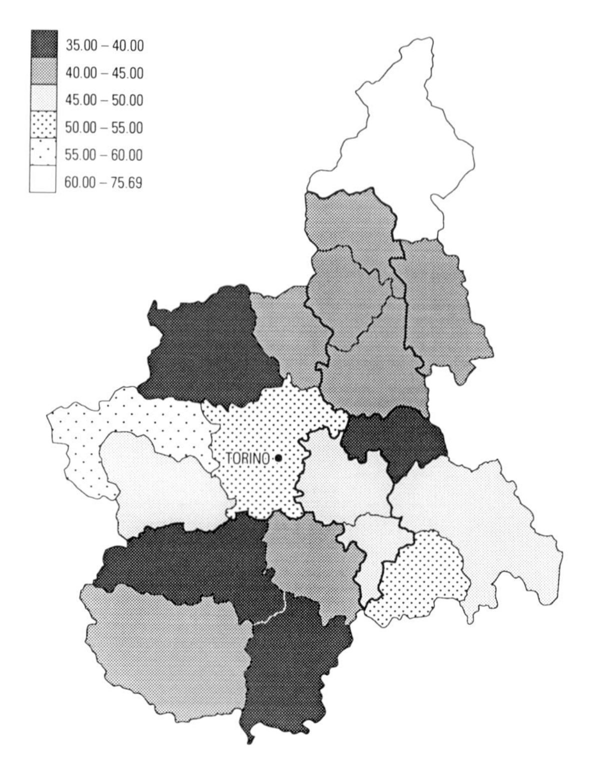

Figure 9.11 Mean (rush hour) travel cost for private transport

Category: transport cost (performance module)
Indicator: mean travel cost (total)

$$C_i = \frac{\Sigma_i F_{ij} C_{ij} + \Sigma_j F_{ij} C_{ij}}{\Sigma_i F_{ij} + \Sigma_j F_{ij}} \tag{9.13}$$

where F_{ij} are flows from zone i to zone j and C_{ij} are rush hour travel times from zone i to zone j

Information: as for (9.1)

Remarks: This kind of indicator has been widely used and tested in transport studies, though generally applied at a smaller spatial scale (urban or metropolitan). It is also possible to distinguish costs perceived either at the origin or at the destination. The problems involved in its construction are

(a) the difficulty of determining to what extent the travel cost is influenced by congestion or the journey length (this problem is intrinsic to the nature of the indicator and can only be resolved by associating it with some measure of the flows generated by the system or of accessibility);

(b) the question (already discussed) of data availability.

It is evident that a more complete and accurate picture could be achieved by constructing a 'battery' of travel cost indicators with specification of trip purpose, transport mode and time period (rush hour, whole day).

9.4 GENERAL REMARKS

In the examples above we have attempted to demonstrate the analytical potential offered by this indicator package. Its application to Piedmont, even though experimental, was able to provide a number of 'snapshots' of the spatial structure of the region.

The pervasiveness of the effects of metropolitan concentration clearly emerged from many of the indicators, i.e. the independence values of industrial activities (Figure 9.8). Some particular features of the region were also highlighted:

1. the relative marginalization of the mountainous northeastern and southeastern areas;
2. the existence of a 'spatial pattern' reflecting both the weight of the metropolitan core and the main transport network (i.e. the influence of public services, Figure 9.7);
3. clear evidence of differences in the wealth ranking of provinces which appear when 'place of production' and 'place of residence' are distinguished (Figure 9.10).

One general point which emerged clearly from the application of the indicators was the importance, especially at smaller scales, of having access to adequate data. This applies not only to the stock variables but also (and perhaps to an even greater extent) to information concerning interactions; the latter is frequently lacking or unsatisfactory.

The examples discussed above show in fact that the validity of an indicator is entirely dependent upon the availability and quality of the necessary data. This is crucial when an exogenous parameter has to be introduced, as the soundness of the estimation procedure and the 'meaningfulness' of the parameter itself also enter into play. (We recall here the problem, discussed above, of the indicators which require a distance impedance parameter.)

While accepting that the fundamental role of an indicator package of this kind is exploratory, there was evidence of a certain paradox. On the one hand, the investigative nature of the package makes it acceptable in principle to compensate for incomplete data input, updating it or generating missing data. There is nothing wrong with producing fictitious matrices or considering a range of alternative parameter values and carrying out a series of sensitivity analyses. This could well produce interesting and useful results. There is also of course the possibility of using modelling techniques, especially spatial interaction models, to provide additional data.

We cannot, on the other hand, altogether ignore the fundamental requirement of indicators, i.e. that of communicability. This is especially important when we are considering the question of their meaningfulness in the context of a particular application. The reliability of the information in this case is crucial.

Finally, in terms of spatial analysis (whether we are speaking of data, performance indicators or models) the one fundamental question we need to ask is if, in the specific context in which they are being applied, they are able to provide us with a real 'information gain' (Nijkamp 1984).

9.5 CONCLUDING COMMENTS

In this chapter we have introduced a package for the calculation of spatial indicators. In order to make an assessment, at this stage a preliminary one, we have looked in turn at the following three aspects:

- conceptual design of the package
- practical features, i.e. flexibility, ease of handling, cost
- computer technology adopted

9.5.1 Conceptual design of the package

As will already be clear from the description, this package can be considered a 'general purpose' tool, with the aim of providing a series of measures of the performance of an urban and regional system. Efforts have been made to respond to practical needs by including the kind of operations most commonly required for the analysis of spatial data.

The design is based on the concept that each indicator is identified with a mathematical 'operator' which transforms the basic data and then presents it in tabular or graphical form. It can therefore be applied to data relating to any aspect of the system, the validity of the 'transformation' being determined by the analyst (user) according to the kind of data involved and the objectives of the analysis. Similarly, the evaluation of the results obtained is left to the interpretive capacity of the analyst.

The concept underlying the kind of indicators used in the PFT package (like those developed in Chapter 5, i.e. residence-based and organization-based) results in a set of indicators 'ready made' for the user, making possible a series of predetermined operations relating to particular subsystems or areas of data. Whether this is an advantage or not remains for the moment difficult to say. Certainly, the general purpose indicators of the RPA package provide greater stimulus to the logical capacity of the user in bringing together concepts and information.

There is no doubt that a combination of the emphasis placed on interactions and the systematic approach shown in Chapter 8 offers considerable potential for the future development of indicators as analytical instruments.

Whether this development should take the form of structurally organized sets of indicators (as in the PFT package) or general purpose indicators (like the ones described here) will depend upon a number of factors including the specific context of the application and the objectives of the analysis. In either case the use of a package of indicators implies two important prerequisites:

1. an analytical capacity on the part of the user, i.e. an understanding of the operations being carried out, a sense of critical awareness and a recognition of the limits and potential of the indicators; it also requires a familiarity with computer technology and information handling (which is still frequently lacking);
2. the availability of relevant data. Information relating in particular to functional and spatial interactions appears to be chronically inadequate (although model-generated information could compensate in part for this deficiency). The carrying out of official and specific surveys and/or encouragement to make existing information more widely available would contribute a great deal to the systematic investigation of the structure of interactions.

9.5.2 Practical features of the package design

It is clear that flexibility, user-friendliness and contained costs are crucial features of any package. From this point of view, the package discussed has certain advantages, which are especially evident in relation to data management. The fact that there is no need for structured or complete data files makes it possible to apply the package to any geographical system and any kind of data (as long as it is logically consistent with the indicator used and does not require modification of the software).

The possibility of 'changing scale' with the automatic reorganization of the data at different levels of zoning (from a finer to a larger resolution) is another extremely useful characteristic.

9.5.3 Computer technology

It is clear that technological progress affecting calculation capacity and computer software will in the future offer even greater potential for the construction and use of packages such as the one developed here, probably at lower cost. We only have to look at the technical improvements achieved in the few years since the PFT package was designed to realize the speed with which innovations are becoming operational.

There remains, however, the sensation that despite the increasing possibilities offered by technology, awareness of the full potential of indicator packages such as this one (which is far more than a simple mapping facility or a data management tool) is still very limited. Whether this is due to their insufficient availability on the market (i.e. lack of supply) or limited awareness on the part of potential users (short-sightedness, scepticism or lack of expertise) are questions which need perhaps to be looked at more closely.

We wish to make two final observations in relation to possible directions for the future development of geographical indicator packages.

First, as mentioned in Chapter 5, it is necessary to develop explicit 'measures' which, based on a set of indicators, are able to express an assessment of the performance of a system, distinguishing

- the performance of a single zone (relative to other zones) and
- the performance of the system as a whole

This is a problem not only in the analysis of single subsystems (residential, retail etc.) and where we wish to distinguish residence-based or organization-based activities but, above all, where we wish to obtain a 'general panorama' of the situation in a system (which, by definition, is spatially defined). Although we are perhaps stepping into the methodological and theoretical area belonging more properly to evaluation techniques, we shall still have to provide indicators with synthesizing ability on which to base the evaluation.

Lastly, it seems likely that the need for synthesizing indicators at both zonal and system level will orient development towards the design of packages which are relatively independent but at the same time easily interfaced.

10 Concluding comments

C.S. Bertuglia, G.P. Clarke and A.G. Wilson

A fundamental concern of urban geography is how well residents are served by the city and how variations in well-being can be effectively monitored and evaluated. Traditional indicators of deprivation or service delivery are typically based on single areas and on published data, and are thus subsequently limited. The central aim of this book has been to examine the changing role of urban models with respect to both the need to re-address measures of urban well-being and the perceived need to make model outputs more in tune with key planning problems of the day. In particular, it has been argued that, whilst there has been substantial progress with a wide range of theoretical problems in urban modelling, modellers have not paid enough attention to the usefulness of their model outputs in terms of indicators which offer new insights into the workings of the city or region. Too often in the past modellers have focused on the direct use of model predictions for planning without fully exploring the rich information base created from both data input into models and the outputs of model simulations. The concluding comments which follow offer some thoughts on a possible research agenda to develop further the ideas put forward in this collection.

It has also been argued that relatively little progress has been made in the past with the use of performance indicators in planning. This may have been largely because of a failure in communication: those social scientists working on indicators have been largely unconnected with modelling; and modellers have tended to work directly with model outputs. The crucial links come through the use of the various *interaction arrays* which are computed within models. They allow the notions of area indicators to be extended considerably. There are two main consequences of providing the idea of performance indicators with a sounder conceptual base in this way. First, better area indicators can be calculated, particularly through the concepts of 'notional provision' and 'catchment population'. Second, the spatial distribution of area indicators can be understood better in relation to connections of any one area to the rest of the system. We are thus offering a new geography of performance indicators for the public and private sector based on the principles of spatial interaction.

One key issue that has emerged is the need to persuade urban modellers to make the outputs of their research more useful to applied planning problems of contemporary societies, both in terms of private sector and public sector issues and in relation to whatever sort of model (static, dynamic, integrated, single sector). The argument of the book has been to offer a framework for defining suites or batteries of performance indicators which may be related to either residential areas (households and individuals) or facility locations (workplaces, shopping centres, schools, hospitals). In this way, it is possible to shed new light on the age-old geographical issues of equity and efficiency. The former is likely to be related to how well served residents of a city or region are in relation to jobs and services. In the case of efficiency the key questions relate to how well organizations (of all types) 'perform' in relation to their potential market.

A second key requirement is the need to build much more experience with the routine calculation of large numbers of performance indicators and the interpretations of these in planning contexts. This experience will almost certainly lead both to a clearer focus on particular indicators in relation to urban and regional problems and to an expansion of the set of indicators as new needs become clear. In fact, although we have started this process in Chapters 8 and 9, what is now required is a number of ways to move forward. First, we need new research programmes to extend the range of indicators to cover additional types of application areas and additional types of model. For example, Occelli (1989) has undertaken a feasibility study for the inclusion of *environmental indicators* into a regional geographic information system. The basic questions here are whether we can identify and investigate the interdependence between the socio-economic systems (outlined throughout this book) and environmental systems. Are we able to define the nature and relevance of environmental problems and the meaning of 'quality' in both an objective and a subjective sense? Can we identify the range of possible policies available to tackle the problem and evaluate their effectiveness? Building on the work of Vos *et al.* (1985) Occelli (1989) has begun this process by considering a wide range of indicators relating to the natural environment. This is very much a first attempt to quantify the link between the strictly environmental indicators (pollution, resources etc.) and those relating to the *use* of resources (socio-economic activities) discussed in detail here.

A second way to make progress is to identify research programmes aimed at quantifying and monitoring the battery of performance indicators for an entire city or region, or better still for a range of cities or regions across the globe. Provided an agreed set of indicators could be defined (which are all measurable or possible to calculate for whichever spatial region) then we could find new ways of comparing the performance of world cities other than the standard economic and social indicators outlined in Chapter 2.

A third broad research area would be to compare the results from such a model-based performance indicators study of quality of life with those based on other research disciplines in human geography and related social sciences.

For example, what could such an indicator-based study offer to the debate on 'localities' exemplified by Cooke (1989). Many of the indicators used within such studies are clearly dominated by the standard published data sets available on a macro basis. There are exceptions to this and one might argue that combining the types of model-based outputs with the incisive qualitative study of Beynon *et al.* (on Teeside) would yield fertile grounds for future collaborative research. That may sound over-ambitious but it leads on to an important consideration of how such indicator studies are best used.

As with any kind of evaluation procedure, there will be two main kinds of uses of a performance indicator system of this kind. First, it will be an aid in the identification of problems and it will provide the information which is the basis for making judgements about which are the most serious. This kind of analysis will help in the definition of projects which will contribute to the solution of problems. Second, the system can be used to evaluate *alternative* projects.

The future research priorities can be summarized in broad terms as follows. First, a wide range of performance indicators should be calculated from model outputs as a matter of routine so that good profiles can be constructed based on population groups, economic or service sectors or areas within a city or region. Second, it should be possible to be more inventive in defining new performance indicators – for example from the implied *utilities* in different submodels and from new submodels such as those based on the environment. In addition, it would be instructive to extend models to focus more explicitly on income and other money flows. This would facilitate the definition of useful new performance indicators. Third, it would be fruitful to investigate further the concepts of 'role' and 'balance' in relation to population, sector and areas: i.e. what is the function of an area in terms of its ability to supply jobs, homes, services and infrastructures to its population, and how do areas interact with each other in terms of inflows and outflows (be they journeys to work, schools, shops, hospitals etc.). Fourth, performance indicators should be related to, and interpreted in terms of, planning problems and ways of solving them. At this point, it is particularly important to recall the systemic basis from which the performance indicators were constructed.

Finally, we recall an earlier comment about the nature and role of performance indicators: that they should be taken as preliminary indications of problems, and that deeper analyses will often be required. There may be opportunities to use them to build bridges between model-based analyses and other approaches to theory and explanation, or at least to begin negotiations! If we could get that far then exciting times lie ahead for urban modelling and we hope for urban geography as a whole.

Appendix An integrated urban model

C.S. Bertuglia, S. Occelli, G.A. Rabino and R. Tadei

1 INTRODUCTION

The integrated model developed in Bertuglia *et al.* (1990b) is based on three main approaches:

1. the master equation approach, which provides the general accounting framework for population mobility;
2. the stock dynamics approach for modelling the economic structures;
3. both economic theory and stochastic extremal processes for deriving specifications for the behavioural models and price formation mechanisms.

The three main 'classic' subsystems of an urban activity are fully dealt with in the model:

- the housing market
- the labour market
- services

Explicit treatments of the land market and transport systems are excluded. While this can be justified for the land market on the grounds of the present operationalization capability, it might prove a serious shortcoming for the transport system. Many transport variables are in fact embedded in the above subsystems; what is missing is the dynamics of stocks and prices in the transport sector which had to be omitted or taken exogenously.

The medium-term time scale is emphasized, focusing on processes which undergo changes over a number of years, such as housing mobility, labour mobility and locational changes in services. Although there are many drawbacks to this assumption, it can help to justify treating as constants some very slowly varying quantities, such as total population or the housing stock.

In terms of actual computation, either discrete or continuous time can be used. Only the former is considered here. An outline of continuous-time treatment is presented in Bertuglia *et al.* (1990a).

In the following discussion, we first recall the general accounting framework of the model and then illustrate the various model versions which have been derived.

2 GENERAL ACCOUNTING FRAMEWORK

i, j, k — the subscripts (or superscripts) labelling zones, $i, j, k = 1, \ldots, n$. In this notation, a superscript refers to the new residential location (unless it is zero, when it indicates a shift to unemployment)

t, Δ — the time indices, where Δ is a small time interval (i.e. one year), $t = 0, \Delta, 2\Delta$

$q_{ij}^{k}(t, \Delta)$ — the probability that people living in zone i and working in zone j at time t move to a dwelling in zone k during the period $(t, t + \Delta)$

$q_{i0}^{k}(t, \Delta)$ — the probability that an unemployed person, living in zone i at time t, moves to a dwelling in zone k during the period $(t, t + \Delta)$

$p_{ij}^{k}(t, \Delta)$ — the probability that a workplace in zone j, occupied by a worker living in zone i at time t, is occupied by a worker living in zone k at time $t + \Delta$

$p_{ij}^{0}(t, \Delta)$ — the probability that a workplace in zone j, occupied by a worker living in zone i at time t, is closed down during the period $t + \Delta$

$\rho_{ij}(t, \Delta)$ — the probability that a new workplace is opened in zone j and occupied by a worker living in i; this probability applies to a stock of potential workplaces which are exogenously given

P — total population, which is assumed to be exogenously given, i.e. no births, deaths or migrations to and from the rest of the world occur in the area:

$$P(t) = \sum_{i=1}^{n} \sum_{j=0}^{n} P_{ij}(t)$$

Q_i — total housing stock in zone i, which is also exogenous for each zone (i.e. no new constructions or demolitions take place)

W_i — vacant dwellings in zone i:

$$W_i = Q_i(t) - \sum_{j=0}^{n} P_{ij}(t)$$

S_j — the maximum number of possible workplaces in zone j, which is assumed to be exogenously specified from land-use constraints

$T_j(t)$ — the number of possible workplaces which can be opened in zone j at time t:

$$T_j(t) = S_j(t) - \sum_{i=1}^{n} P_{ij}(t)$$

L_i total labour demand for workers living in zone i

$\delta_1, \delta_2, \beta_1 > 0$ parameters

The difference equations accounting for all the changes of state of the employed and unemployed population are respectively

number in a state at time $t + \Delta$	number in a state at time t	residential moves into i for workers currently employed in j	all currently unemployed in i who find jobs in j at the expense of those living in k

$$P_{ij}(t+\Delta) = P_{ij}(t) + \sum_{k=1}^{n} P_{kj}(t)\left[q^{i}_{kj}(t,\Delta) + p^{i}_{kj}(t,\Delta)\right]$$

	residential moves from i to any k	loss of job at j and replacement by unemployed living anywhere	the component arising from new workplaces being created

$$- P_{ij}(t)\left[\sum_{k=1}^{n} q^{k}_{ij}(t,\Delta) + \sum_{k=0}^{n} p^{k}_{ij}(t,\Delta)\right] + T_j(t)\rho_{ij}(t,\Delta) \quad (1)$$

number in a state at time $t + \Delta$	number in a state at time t	residential moves into i for workers currently unemployed	residential moves out of i for workers currently unemployed

$$P_{i0}(t+\Delta) = P_{i0}(t) + \sum_{k=1}^{n} P_{k0}(t) q^{i}_{k0}(t,\Delta) - P_{i0}(t) \times \sum_{k=1}^{n} \frac{q^{i}_{k0}(t,\Delta)}{q^{k}_{i0}}$$

		workers living in i who become unemployed	unemployed in i who become employed as a result of job replacement

$$+ \sum_{j=1}^{n} P_{ij}(t) \sum_{k=0}^{n} p^{k}_{ij}(t,\Delta) - \sum_{k=1}^{n} \sum_{j=1}^{n} P_{kj}(t) \times p^{i}_{kj}(t,\Delta)$$

$$- \sum_{j=1}^{n} T_j(t)\rho_{ij}(t,\Delta) + \delta_1 \frac{L_i(t,\Delta)}{P_{i0}(t)} \times W_i(t)\, q^{0}_{i}(t,\Delta)$$

$$- \delta_2 \left[P_{i0}(t) - L_i(t,\Delta)\right] \quad (2)$$

and

$$q_i^0(t, \Delta) = \frac{W_i(t)\ \exp[\beta_1 u_{i0}(t)]}{\Sigma_{k=1}^n W_k(t)\ \exp[\beta_1 u_{i0}(t)]}$$

Summing equations (1) and (2) over all zones yields, with some rearrangement,

$$P(t+\Delta) = P(t) \qquad \text{for all } t$$

This provides one accounting basis for the model (see Chapter 6 for alternatives).

Alternative model versions can be developed according to the different specifications of the transition probabilities – which have a logit type formulation – in relation to different behavioural hypotheses. Three model versions were derived:

Model A — transition probabilities are specified according to residence-focused market clearing

Model B — transition probabilities are specified according to workplace market clearing

Model C — more realistic assumptions are introduced into the specifications of transition probabilities

3 MODEL A

The transition probabilities $q_{ij}^k(t, \Delta)$, $p_{ij}^k(t, \Delta)$ and $p_{ij}^0(t, \Delta)$ for the

$$q_{ij}^k(t, \Delta) = \frac{\lambda \Delta W_k(t)\ \exp[\beta_1 u_{kj}(t) + \beta_2 v_{ij}(t)]}{D_j^i(t, \Delta)} \tag{3}$$

$$p_{ij}^k(t, \Delta) = \frac{\mu \Delta P_{k0}(t)\ \exp[\beta_1 u_{ij}(t) + \beta_2 v_{kj}(t)]}{D_j^i(t, \Delta)} \tag{4}$$

$$p_{ij}^0(t, \Delta) = \frac{\gamma \Delta\ \exp[\beta_1 u_{ij}(t)]}{D_j^i(t, \Delta)} \tag{5}$$

employed are expressed respectively by
and the denominator is given by

$$\begin{aligned} D_j^i(t, \Delta) = {} & \lambda \Delta \exp[\beta_2 v_{ij}(t)] \sum_{k'=1}^{n} W_{k'}(t) \exp[\beta_1 u_{k'j}(t)] \\ & + \Delta \exp[\beta_1 u_{ij}(t)] \left\{ \mu \sum_{k'=1}^{n} P_{k'0}(t) \exp[\beta_2 v_{k'j}(t)] + \gamma \right\} \\ & + \exp[\beta_1 u_{ij}(t) + \beta_2 v_{ij}(t)] \end{aligned} \tag{6}$$

where $u_{ij}(t)$ is the utility function of a dwelling in zone i for people working in zone j, and is defined as

$$u_{ij}(t) = -[c_{ij} + \bar{c}_i(t) + r_i(t)]$$

assuming that housing choice is related to a minimization of total cost and deriving from the sum of

c_{ij}, the cost of a return trip between zones i and j,
$\bar{c}_i$, the average per capita expenditure on services for people living in zone i, and
$r_i(t)$, the average dwelling price in zone i.

$v_{ij}(t)$ is the utility function for a firm in j hiring a worker from zone i (i.e. the profit). $v_{ij}(t)$ is defined as

$$v_{ij}(t) = d_j(t) - [w_{ij}(t)]$$

which means that the worker choice by firms is related to a maximization of total profit given by, neglecting other costs, the difference between

$d_j(t)$, the average revenue per unit of employment at a workplace in zone j, and
$w_{ij}(t)$, the wage a firm in zone j pays for a worker from zone i:

$$w_{ij}(t) = [c_{ij} + \bar{c}_i(t) + r_i(t)] + y_i(t)$$

where $y_i(t)$ is the disposable income for people living in zone i.

λ is a non-negative parameter representing a rate of diffusion of information about vacant dwellings available; β_1, β_2 are non-negative parameters; and μ is a parameter measuring the intensity of hiring and firing decisions.

Equations (3)–(5) are in fact the result of the combination of two distinct processes whose mathematical formalization is omitted here for brevity (see Bertuglia *et al.* 1990b, ch. 17):

1 a housing change decision, specified without considering labour changes;
2 a job change decision, specified without considering housing changes.

The simultaneous occurrence of the above processes is also excluded here. It is assumed that each individual can undergo at most one change during the period $(t, t + \Delta)$.

The transition probabilities $q_{i0}^k(t, \Delta)$, $\rho_{ij}^k(t, \Delta)$ for the unemployed are expressed by

$$q_{i0}^{k}(t, \Delta) = \frac{\lambda \Delta W_k(t) \exp(\beta_1 u_{k0})}{\lambda \Delta \Sigma_{k=1}^{n} W_k(t) \exp[\beta_1 u_{k0}(t)] + \exp[\beta_2 v_{ij}(t)] + 1} \quad (7)$$

$$\rho_{ij}(t, \Delta) = \frac{\mu \Delta P_{i0}(t) \exp[\beta_2 v_{ij}(t)]}{\mu \Delta \Sigma_{k=1}^{n} P_{i0}(t) \exp[\beta_2 v_k(t)] + 1} \quad (8)$$

where the utility function for the residential choice reduces to:

$$u_{i0}(t) = -[\bar{c}_i(t) + r_i(t)]$$

Equations (3)–(8) specify all the transition probabilities required in the difference equations (1) and (2). It can easily be checked that they are consistent probabilities. In particular:

$$\sum_{k=1}^{n} [q_{ij}^{k}(t, \Delta) + p_{ij}^{k}(t, \Delta)] + p_{ij}^{0}(t, \Delta) < 1 \quad (9)$$

$$\sum_{k=1}^{n} q_{i0}^{k}(t, \Delta) < 1 \quad (10)$$

$$\sum_{i=1}^{n} \rho_{ij}(t, \Delta) < 1 \quad (11)$$

state the conservation properties of the system. The total probability of a change of state in $(t, t+\Delta)$, for any initial state, is less than 1. Its complement to unity is of course the probability of making no change.

The computation of the time-varying signals entering in the definition of the utility functions used in the transition probabilities is based on the following assumptions.

1. All the markets – housing, labour and services – are cleared within one period. Δ is assumed to be large enough for the demand–supply interactions to settle down to partial equilibrium at the end of each period.
2. The price adjustment process is faster than changes in quantities (population, economic activity levels, stocks) and its dynamics can be considered to be very short term.
3. Demand behaviour is still described by logit models, while supply behaviour is assumed to be in accordance with bid-price maximizing.
4. For the service sector in particular, a modified flow model is used, where no sector disaggregation is retained. In addition, demand for services is considered to be in equilibrium at every instant in time.

Housing market clearing

Using equations (3) and (7), the total demand for housing in zone i, $H_i(t, \Delta)$, is obtained as

$$H_i(t, \Delta) = \sum_{k=1}^{n} \sum_{j=0}^{n} P_{kj}(t)\, q^i_{kj}(t, \Delta) \tag{12}$$

The total supply of housing in zone i in $(t, t+\Delta)$ is given by

$$W_i(t)\{1 - \exp[-\phi_i(t, \Delta)]\} \tag{13}$$

where the expression in the braces represents the probability that a landlord owning a vacant dwelling is willing to sell if he/she has received at least one non-negative bid. $\phi_i(, \Delta)$ is defined as

$$\phi_i(t, \Delta) = \frac{\exp[\beta_1 r_i(t)]\ H_i(t, \Delta)}{\Delta W_i(t)}$$

The market clearing equations for house prices $r_i(t)$ are

$$H_i(t, \Delta) = W_i(t)\{1 - \exp[-\phi_i(t, \Delta)]\} \qquad i = 1, \ldots, n \tag{14}$$

Labour market clearing

Using equations (4) and (8), the total demand for workers living in zone i, $L_i(t, \Delta)$, is given by

$$L_i(t, \Delta) = \sum_{j=1}^{n} \sum_{k=1}^{n} P_{kj}(t)\, p^i_{kj}(t, \Delta) + \sum_{j=1}^{n} T_j(t)\, \rho_{ij}(t, \Delta) \tag{15}$$

The total supply of labour from the unemployed living in zone i in period $(t, t+\Delta)$ is given by

$$P_{i0}(t)\{1 - \exp[-\psi_i(t, \Delta)]\} \tag{16}$$

where $\psi_i(t, \Delta)$ is defined as

$$\psi_i(t, \Delta) = \frac{\exp[\beta_2 y_i(t)]\ L_i(t, \Delta)}{\Delta P_{i0}(t)} \qquad i = 1, \ldots, n$$

This means that applicants offered jobs are assumed to be disposable-income maximizers: they evaluate alternative jobs according to the disposable income they provide. Once fixed consumptions are deducted they choose the one providing the highest disposable income. Otherwise, they prefer to remain unemployed and not to contribute to the labour supply.

The market clearing equations for the disposable incomes $y_i(t)$ are

$$L_i(t, \Delta) = P_{i0}(t)\{1 - \exp[-\psi_i(t, \Delta)]\} \qquad i = 1, \ldots, n \tag{17}$$

Service market clearing

The total demand for services in j at time t, $\Sigma_i F_{ij}(t)$, is provided by the following flow model:

$$F_{ij}(t) = G_i(t) \frac{A_j^{\alpha}(t) \exp\{-\beta_3[c_{ij} + x_j(t)]\}}{\Sigma_{j=1}^{n} A_j^{\alpha}(t) \exp\{-\beta_3[(c_{ij} + x_j(t)]\}} \quad (18)$$

Here β_3 and α are non-negative parameters and $G_i(t)$ is the total flow of customers from zone i, $i = 1, \ldots, n$:

$$G_i(t) = \psi_{1i} \sum_{k=0}^{n} P_{ik}(t) + \psi_{2i} \sum_{k=1}^{n} P_{ki}(t)$$

ψ_{1i}, ψ_{2i} are the frequencies of trips generated from residences and workplaces, respectively. $A_j(t)$ is the total size of service activities in zone j, measured in terms of employment:

$$A_j(t) = \sum_{k=1}^{n} P_{kj}(t)$$

$x_j(t)$ is the average price for an average unit of service offered in zone j.

From equation (18) the revenues $d_j(t)$ can be computed as

$$d_j(t) = -\frac{\delta}{\beta_3} \ln \left\{ \frac{\Sigma_{j=1}^{n} A_j^{\alpha}(t) \exp[-\beta_3(c_{ij} + x_j)]}{\Sigma_{j=1}^{n} A_j^{\alpha}(t)} \right\} \qquad j = 1, \ldots, n \quad (19)$$

where $\delta > 0$ is a parameter and $\theta_j(t)$ is defined as

$$\theta_j(t) = \frac{\exp[\beta_3 x_j(t)] F_j(t)}{a A_j^{\alpha}(t)} \qquad j = 1, \ldots, n$$

The average per capita expenditure in service consumption, $\bar{c}_i(t)$, for people living in zone i is computed as

$$\bar{c}_i(t) = -\frac{\delta}{\beta_3} \ln \left\{ \frac{\Sigma_{j=1}^{n} A_j^{\alpha}(t) \exp[-\beta_3(c_{ij} + x_j)]}{\Sigma_{j=1}^{n} A_j^{\alpha}(t)} \right\} \qquad j = 1, \ldots, n \quad (20)$$

The market clearing equations for the service prices, $x_j(t)$, are

$$F_j(t) = a A_j^{\alpha}(t) \{1 - \exp[-\theta_j(t)]\}$$

where a is the number of customers and

$$F_j(t) = \sum_{i=1}^{n} F_{ij}(t)$$

4 MODEL B

The accounting structure and the definition of the main variables are the same as in model A. The main difference is the way in which the labour market is assumed to operate. Market clearing is focused on each employment zone, generating a set of wage rates w_{jt}. This produces a marginal change in the residential model and deeper changes in the labour market model.

The equations of the transitional probabilities $q_{ij}^k(t, \Delta)$, $p_{ij}^k(t, \Delta)$, $p_{ij}^0(t, \Delta)$ for the unemployed remain the same. The only modifications concern the following.

1 The definition of the utility function, in which wage rates enter as a new component

$$u_{ij}(t) = w_j(t) - [c_{ij} + \bar{c}_i(t) + r_i(t)]$$

In this version $u_{ij}(t)$ becomes disposable income and takes into account both wage and house prices as well as the transport and service elements of the cost of living in i.

2 The definition of the utility function for a firm, where $v_{ij}(t)$ is substituted by $v_j(t)$, as

$$v_j(t) = d_j(t) - w_j(t)$$

and $w_j(t)$ is expressed by

$$w_j(t) = [c_{ij} + \bar{c}_i(t) + r_i(t)] + y_i(t)$$

For the unemployed, the transitional probability $q_{i0}^k(t, \Delta)$ remains as before and the utility function for the residential choice becomes

$$u_{i0}(t) = w_0(t) - [\bar{c}_i(t) + v_i(t)]$$

where $w_0(t)$ is the total benefit paid to an unemployed person.

$\rho_{ij}(t, \Delta)$ takes the following new form:

$$\rho_{ij}(t) = \frac{P_{i0}(t) \exp[\beta_4 u_{ij}(t)]}{\Sigma_k P_{k0}(t) \exp[\beta_4 u_{kj}(t)]} \frac{\bar{\mu}\Delta \exp[\beta_2 v_j(t)]}{\Sigma_k \exp[\beta_2 v_k(t)]} \qquad (21)$$

where β_4 is a parameter relating to the workplace choices of newly hired workers and $\bar{\mu}$ is a new constant relating to the probability of a new workplace being opened.

The housing and services market clearings remain the same as in model A. The alternative version for labour market clearing is as follows. The total number of vacant jobs at t in zone j, $v_j(t, \Delta)$ is defined as

$$v_j(t, \Delta) = E_j(t) \sum_{k=1}^{n} \bar{p}_{ij}^{k}(t, \Delta) + \sum_{i=1}^{n} T_j(t)\, \rho_{ij}(t, \Delta) \tag{22}$$

where $E_j(t) = \Sigma_i P_{ij}(t)$ is the total number of jobs in j at time t and $\bar{p}_{ij}^{k}(t, \Delta)$ is the probability that a workplace in zone j occupied by a worker living in zone i at time t is occupied by a worker living in zone k at time $t + \Delta$ (defined without considering housing change):

$$\bar{p}_{ij}^{k}(t, \Delta) = \frac{\mu \Delta P_{k0}(t) \exp[\beta_2 v_j(t)]}{\mu \Delta \{1 + \exp[\beta_2 v_j(t)] \, \Sigma_{k'=1}^{n} P_{k'0}(t)\} + \exp[\beta_2 v_j(t)]}$$

$\Sigma_i T_j(t)\, \rho_{ij}(t, \Delta)$ is the number of new jobs.

The number of vacancies filled by workers resident in i can be assumed to be

$$\Pi_{ij}(t, \Delta) = \frac{P_{i0}(t) V_j(t) \exp[\beta_4 v_{ij}(t)]}{\Sigma_k V_k(t) \exp[\beta_4 v_{ik}(t)]} \tag{23}$$

The market clearing equation now has a similar form to those for housing and services and becomes

$$V_j(t) = \sum_{i} \Pi_{ij}(t) \{1 - \exp[-\psi_j(t)]\} \tag{24}$$

where

$$\psi_j(t) = \exp[\beta_2 w_j(t)] \frac{V_j(t)}{\Sigma_i \Pi_{ij}(t)}$$

is the average time of potential demand for workers in i.

Note that the term in square brackets in equation (24) represents the probability that a workplace in j is taken by any worker independently of his residence in i.

The difference from model A is that now the independent variable is the state function w_j, the wage in j, not y_i, the disposable income in i. Wages are arrived at through a negotiation process between workers and firms. The latter are not aware of the detailed consumption list of the workers. Workers, however, know the composition of their budget constraints and negotiate their wage with the firms in such a way as to get the highest possible value for their disposable income. This is a result of a maximization the workers carry out over a whole list of jobs.

5 MODEL C

This reformulated version of the model considers:

1 a notation for the transition matrix which unifies the p and q matrices, allowing more general transitions to occur;
2 a theoretical model of the labour market which is underpinned by a discrete choice process in which workers select jobs in a market setting;
3 some relaxation of the restrictive dynamic assumptions relating to population and employment stocks.

This model version also uses simplified notational definitions.

P	the total population in an area
Q_i	the total housing stock in zone i
S_j	the maximum number of possible workplaces in zone j
$T_{ij}(t)$	the number of people living in i and working in j at time t
$E_j(t)$	the number of new workplaces which can be opened in zone j at time t with $E_j(t) = S_j - \sum_{i=1}^{n} T_{ij}(t)$
$T_{i0}(t)$	the number of unemployed people living in i at time t
$q_{ij}^{k}(t, \Delta)$	the probability that people living in zone i and working in zone j at time t will move to a dwelling in zone k during the period $(t,\ t + \Delta)$; hereafter the assumption will be made that all transition rates refer to the interval $(t, t + \Delta)$
$q_{i0}^{k}(t, \Delta)$	the probability that an unemployed person living in zone i will move to a dwelling in zone k
$p_{ij}^{k}(t, \Delta)$	the probability that a workplace in zone j occupied by a worker living in i at time t will be occupied by a worker living in k at time $t + \Delta$
$p_{ij}^{0}(t, \Delta)$	the probability that a workplace in zone j occupied by a worker living in i is closed down
$\rho_{ij}(t, \Delta)$	the probability that a new workplace is opened in zone j and occupied by a worker living in i

The new accounting equations for the employed and unemployed population become, respectively,

$$T_{ij}(t+\Delta) = T_{ij}(t)\left(1 - \sum_{k=1}^{n} q_{ij}^{k} - \sum_{k=0}^{n} p_{ij}^{k}\right) + \sum_{k=1}^{n} T_{kj}(t)\left(q_{kj}^{i} + p_{kj}^{i}\right) + E_j(t)\rho_{ij}(t, \Delta) \quad (25)$$

$$T_{i0}(t+\Delta) = T_{i0}(t)\left[1 - \sum_k q^k_{i0}\right] + \sum_k T_{k0}(t)\, q^i_{k0}(t,\Delta)$$
$$+ \sum_j T_{ij}(t) \sum_k p^k_{ij}(t,\Delta) - \sum_k \sum_j T_{kj}(t)\, p^i_{kj}(t,\Delta)$$
$$- E_j(t)\rho_{ij}(t,\Delta) \qquad (26)$$

Transition probabilities

The probability of making a transition between the states (i, j) to (k, l) is given by

$$\text{prob}[(i,j) \Rightarrow (k,l)] = \frac{W(k,l)\ \exp[\beta U(k,l)}{\Sigma_l \Sigma_k W(k,l)\ \exp[\beta U(k,l)]} \qquad (27)$$

The transitions to states which involve a change in only one label, and are therefore confined to either the housing or the labour market, follow immediately. Of course, the probability of remaining in the present state is $\text{prob}[(i, j) \Rightarrow (i, j)]$ or 1 minus the probability of transitions to all other states.

In this model version the original restriction that mobility in the labour market arises from unemployed workers taking the jobs of workers who have been fired is removed. The q^k_{ij} and p^k_{ij} probabilities are then given by

$$q^k_{ij} = \frac{\lambda \Delta Q_k(t)\ \exp[\beta(u_{kj} + v_{ij})]}{D_{ij}} \qquad (28)$$

$$p^k_{ij} = \frac{\lambda \Delta P(t)\ \exp[\beta(u_{ij} + v_{kj})]}{D_{ij}} \qquad (29)$$

in which the denominator D_{ij} is given by

$$D_{ij} = \exp[\beta(u_{ij} + v_{kj})] + \Delta\left\{\exp(\beta v_{ij}) \sum_{k'} Q_{k'} \exp(\beta u_{k'j})\right.$$
$$\left. + \mu \exp(\beta u_{ij})\left[\sum_{k'} P \exp(\beta v_{k'j} + 1)\right]\right\} \qquad (30)$$

where u_{ij} is the residential utility function as in the previous versions and v_{ij} is the net utility of choosing a job in zone j (whose definition derives from a reformulation of the labour market which is dealt with in the following).

The reformulated labour market

In the reformulated labour market model it is assumed that

1 individuals select a job from a discrete set ($j = 1, \ldots, n$) on the basis of

maximum utility (or, alternatively, maximum disposable income);
2 in any given period an individual will either work or be unemployed;
3 in a given period a fixed stock of jobs exists;
4 wages will be determined through the usual clearing assumptions which equate demand to supply.

The net (representative) utility of choosing a job in zone j (given a residence in zone i) is

$$v_{ij} = U_{jw} - c_{ij} - C_i - r_i \tag{31}$$

where U_{jw} is the maximum utility derived from a job in zone j offering (an average zonal) wage w_j, c_{ij} is the cost of a return trip between zones i and j, C_i is the total cost of service trips from the residential zone i and r_i is the residential rent in zone i (which is given in the labour market model).

Introducing a random term accounting for dispersion of preferences, zonal aggregation effects etc., the probability p_{ij} that an individual will select state j is

$$p_{ij} = \frac{W_j \exp(\lambda v_{ij})}{\Sigma_j W_j \exp(\lambda v_{ij})} \tag{32}$$

where λ is a parameter accounting for random effects and W_j is the number of jobs in zone j.

Market clearing is achieved and an equilibrium wage distribution over the zones is determined by equating the supply of labour to a zone, as a function of the relative prices, with the number of jobs W_j:

$$\sum_i P_i p_{ij} \leq W_j \tag{33}$$

To provide a link with the previous versions of the model it is established that $\exp(\psi_j)$ represents the probability that a supply unit remains vacant. Equation (33) then becomes

$$\sum_i P_i \frac{W_j \exp(\lambda v_{ij})}{\Sigma_j W_j \exp(\lambda v_{ij})} = W_j[1 - \exp(-\psi_j)] \tag{34}$$

Market clearing

As in the previous model the total demand for and supply of houses in zone k are defined respectively as

$$H_i(t, \Delta) = \sum_{k=1}^{n} \sum_{k=0}^{n} P_{kj}(t)\, q^i_{kj}(t, \Delta) \tag{35}$$

$$Q_i(t)\{1 - \exp[-\phi_i(t, \Delta)]\} \tag{36}$$

The total supply of labour to zone j is similarly

$$L_j(t, \Delta) = \sum_i P_i(t) p_{ij}(t, \Delta) \tag{37}$$

and the total supply of jobs will vary according to which model is considered.

In either model, the current version or the one including the $1 - \exp(-\phi_j)$ terms, two equations for demand and supply of housing and labour must be solved but each involves a lagged expression for the price (housing rent or wage) in the other market.

In this model there exists in effect a single 'market' for residences and jobs within which combinations of options are sought and prices are determined (at the relevant location) to equate supply and demand.

References

Allen, P.M. (1982) 'Evolution, modelling and design in a complex world', *Environment and Planning B, 9*, 95–111.

Allen, P.M. and Sanglier, M. (1979) 'A dynamic model of growth in a central place system', *Geographical Analysis, 11*, 256–72.

Allen, P.M. and Sanglier, M. (1981) 'Urban evolution, self-organisation and decision-making', *Environment and Planning A, 13*, 167–83.

Allen, P.M., Deneubourg, J.L., Sanglier, M., Boon, F. and de Palma, A. (1978) *The Dynamics of Urban Evolution*, vol. 1: *Interurban Evolution*, vol. 2: *Intraurban Evolution*, University of Bruxelles, Final Report prepared for US Department of Transportation, Research and Special Programs Administration, Washington, D.C.

Alonso, W. (1964) *Location and Land Use*, Harvard University Press, Harvard, Mass.

Amos, J.C. (1970) 'Social malaise in Liverpool: interim report on social problems and their distribution', City of Liverpool.

Anas, A. (1982) *Residential Location Models and Urban Transportation*, Academic Press, London.

Anas, A. (1983) *The Chicago Area Transportation–Land Use Analysis System*, NorthWest University, Evanston, Ill.

Anas, A. (1987) *Modelling in Urban and Regional Economics*, Vol. 25, Harwood, New York.

Anderson, J.G. (1972) 'Model of a health service system', *Health Services Research, 7*, 23–42.

Anderson, J.G. (1973) 'Causal models and social indicators', *American Sociological Review, 38*, 285–301.

Andrews, F.M. (1981) 'Subjective social indicators, objective social indicators, and social accounting systems', in Juster, F.T. and Land, J.C. (eds) *Social Accounting Systems, Essays on the State of the Art*, Academic Press, New York, 377–415.

Angrist, S., Belkin, J. and Wallace, W. (1976) 'Social indicators and urban policy analysis', *Socio-Economic Planning Sciences, 10*, 193–8.

Anson, J. (1991) 'Demographic indices as social indicators', *Environment and Planning A, 23*, 433–46.

Batty, M. (1970) 'An activity allocation model for the Nottinghamshire Derbyshire subregion', *Regional Studies, 4*, 307–32.

Batty, M. (1979) 'Progress, success and failure in urban modelling', *Environment and Planning A, 11*, 863–79.

Batty, M. (1989) 'Urban modelling and planning: reflections, retrodictions and prescriptions', in MacMillan, B. (ed.) *Remodelling Geography*, Basil Blackwell, Oxford, 147–69.

Bauer, R.A. (ed.) (1966) *Social Indicators*, MIT Press, Cambridge, Mass.

Beaumont, J.R. (1987) 'Quantitative methods in the real world: a consultant's view of practice', *Environment and Planning A, 19*, 1441–8.

Beaumont, J.R. (1989) 'Applied geographical modelling: some personal comments', in MacMillan, B. (ed.) *Remodelling Geography*, Basil Blackwell, Oxford, 170–6.

Beckmann, M.J. (1973) 'Equilibrium models of residential location', *Regional and Urban Economics, 3*, 361–8.

Beguinot, C. (1989) *La citta cablata. Un' Enciclopedia*, Giannini, Naples.

Ben-Akiva, M. and Lerman, S.R. (1978) 'Disaggregate and mobility choice models and measures of accessibility', in Hensher, D.A. and Stopher, P.R. (eds) *Behavioural Travel Modelling*, Croom Helm, London, 654–79.

Bennett, R.J. (1985) 'Quantification and relevance', in Johnston, R.J. (ed.) *The Future of Geography*, Methuen, London.

Bennett, R.J. (1989) 'Whither models and geography in a post-welfarist world?', in MacMillan, B. (ed.) *Remodelling Geography*, Basil Blackwell, Oxford, 273–90.

Ben-Shahar, H., Mazor, A. and Pines, D. (1969) 'Town planning and welfare maximization: a methodological approach', *Regional Studies, 3*, 105–13.

Berthoud, R. (1976) 'Where are London's poor?', *Greater London Intelligence Quarterly, 36*, 5–12.

Bertuglia, C.S. (1991) 'La citta' come sistema', in Bertuglia, C.S. and La Bella, A. (eds) *Sistemi Urbani*, vol. 1, *Le Teorie. Il Sistema e le Reti*, Angeli, Milan, 301–90.

Bertuglia, C.S. and La Bella, A. (1991) 'Introduzione', in Bertuglia, C.S. and La Bella, A. (eds) *Sistemi Urbani*, vol. 1, *Le Teorie. Il Sistema e le Reti*, Angeli, Milan, 9–54.

Bertuglia, C.S. and Rabino, G.A. (1975) *Modello per l'Organizzazione di un comprensario*, Guida, Naples.

Bertuglia, C.S. and Rabino, G.A. (1976) 'L'assetto territoriale', in IRES, *Linee di Piano Territoriale per il Comprensorio di Torino*, Guida, Naples, 192–511.

Bertuglia, C.S. and Rabino, G.A. (1990) 'The use of mathematical models in the evaluation of actions in urban planning: conceptual premises and operational problems', *Sistemi Urbani, 12*, 121–32.

Bertuglia, C.S., Leonardi, G., Occelli, S., Rabino, G.A. and Tadei, R. (1987a) 'An integrated model for the dynamic analysis of location–transport interrelations', *European Journal of Operations Research, 31*, 198–208.

Bertuglia, C.S., Leonardi, G., Occelli, S., Rabino, G., Tadei, R. and Wilson, A.G. (eds) (1987b) *Urban Systems: Contemporary Approaches to Modelling*, Croom Helm, London.

Bertuglia, C.S., Rabino, G.A. and Tadei, R. (1988) 'Mathematical model and plan evaluation', *Sistemi Urbani, 10*, 237–61.

Bertuglia, C.S., Cecotti, M., Fabbro, S., Petrossi, F., Piva, F., Propedo, G. and Tadei, R. (1989a) Verso una nuova concezione del riequilibrio territoriale. Una valutazione dei sistemi sub-regionali del Friuli-Venezia giulia secondo criteri di efficacia e di efficienza spaziale, IRES F-VG Papers no. 5, Udine.

Bertuglia, C.S., Rabino, G.A. and Tadei, R. (1989b) 'I modelli matematici e la valutazione dei piani, in Barbanente, A. (ed.) *Metodi di valutazione nella pianificazione urbana e territoriale*, Teoria e casi di studio, Atti del Colloquio Internazionale Capri-Napoli 1988, Quaderno 6 IRIS-CNR, Ragusa Grafica Moderna, Bari, 47–71.

Bertuglia, C.S., Leonardi, G., Occelli, S., Rabino, G.A. and Tadei, R. (1990a) 'A first example of an integrated operational model', in Bertuglia, C.S., Leonardi, G. and Wilson, A.G. (eds) *Urban Dynamics: Designing an Integrated Model*, Routledge, London, 367–94.

Bertuglia, C.S., Leonardi, G. and Wilson, A.G. (eds) (1990b) *Urban Dynamics: Designing an Integrated Model*, Routledge, London.

Bertuglia, C.S., Laurentini, A., Occelli, S. and Piasenza, G. (1991a) 'Indicatori territoriali', ESPI and Regione Siciliona, Palermo, mimeo.

Bertuglia, C.S., Rabino, G.A. and Tadei, R. (1991b) 'La valutazione delle azioni in campo urbano in un contesto caratterizzato dall'impiego di modelli matematici', in Bielli, M. and Reggiani, A. (eds) *Sistemi Spaziali ed Approcci Metodologici*, Angeli, Milan, 97–143.

Bertuglia, C.S., Rabino, G.A. and Tadei, R. (1992) 'Una rassegna delle principali questioni che si pongono oggi nella pianificazione urbana', Quaderni di Scienze Regionali, 4, Dipartimento di Scienze e Tecniche per i Processi di Insediamento, Turin University, Turin.

Bertuglia, C.S., Lombardo, S., Occelli, S. and Rabino, G.A. (1993) 'Innovazioni tecnologiche e trasformazioni territoriali: il modello Telemaco (telematica, localizzazione e mobilità: analisi e controllo' in PFT2, *Atti del 1º Convegno Nazionale*, Rome, 879–910.

Birkin, M. (1987) 'An iterative proportional fitting procedure', *Computer Manual 26*, School of Geography, University of Leeds.

Birkin, M. and Clarke, G.P. (1987) 'Synthetic data generation and the evaluation of urban performance', Working Paper 502, School of Geography, University of Leeds.

Birkin, M. and Clarke, G.P. (1992) 'Towards effective decision support systems for the planning of training provision', *British Journal of Education and Work, 5 (3)*, 79--91.

Birkin, M. and Clarke, M. (1985) 'Comprehensive models and efficient accounting frameworks for urban and regional systems', in Haining, R. and Griffith, D.A. (eds) *Transformations through Space and Time*, Martinus Nijhoff, Dordecht, 165–91.

Birkin, M. and Clarke, M. (1988) 'SYNTHESIS: a synthetic spatial information system with methods and examples', *Environment and Planning A, 20*, 645–71.

Birkin, M. and Clarke, M. (1989) 'The generation of individual and household incomes at the small area level using SYNTHESIS', *Regional Studies, 23*, 535–48.

Birkin, M. and Foulger, F. (1992) 'Sales performance and sales forecasting using spatial interaction modelling: the W H Smith approach', Working Paper 92/21, School of Geography, University of Leeds.

Birkin, M. and Wilson, A.G. (1986a) 'Industrial location models I: a review and an integrating framework', *Environment and Planning A, 18*, 175–205.

Birkin, M. and Wilson, A.G. (1986b) 'Industrial location models II: Weber, Palander, Hotelling and extensions in a new framework', *Environment and Planning A, 18*, 293–306.

Birkin, M., Clarke, G.P., Clarke, M. and Wilson, A.G. (1987) 'Geographical information systems and model-based location analysis: a case of ships in the night or the beginnings of a relationship?', Working Paper 498, School of Geography, University of Leeds.

Birkin, M., Clarke, G.P., Clarke, M. and Wilson, A.G. (1990) 'Elements of a model-based information system for the evaluation of urban policy', in Worrall, L. (ed.) *Information Systems for Urban and Regional Policy Analysis*, Belhaven, London, 133–62.

Birkin, M., Clarke, G.P., Clarke, M. and Wilson, A.G. (forthcoming) *Intelligent GIS*, Longman, London.

Black, J. and Conroy, M. (1977) 'Accessibility measures and the social evaluation of urban structure', *Environment and Planning A, 9*, 1013–31.

Boyce, D.E. and Southworth, F. (1979) 'Quasi-dynamic urban location models with endogenously determined travel costs', *Environment and Planning A, 11*, 575–84.

Bracken, I. (1981) *Urban Planning Methods*, Methuen, London.

Bradford, M.G. (1989) *Educational change in the city*, in Herbert, D.T. and Smith, D.M. (eds) *Social Problems and the City: New Perspectives*, Oxford University Press, Oxford, 142–58.

Bradford, M.G. (1991) 'School performance indicators, the local residential environment and parental choice', *Environment and Planning A, 23*, 319–32.

Brand, J. (1975) The politics of social indicators, *British Journal of Sociology, 26*, 78–90.

Brewer, G.D. (1973) *Politicans, Bureaucrats and the Consultant: a Critique of Urban Problem Solving*, Basic Books, London.

Burns, L.D. (1979) *Transportation, Temporal and Spatial Components of Accessibility*, Lexington Books, Lexington, Mass.

Burtenshaw, D., Bateman, M. and Ashworth, G.J. (1991) *The European City: a Western Perspective*, David Fulton, London.

Camagni, R. (1988) 'Lo spazio della pianificazione', in Gibelli, M.C. and Magnani, I. (eds) *La Pianificazione Urbanistica come Strumento di Politica Economica*, Angeli, Milan, 61–71.

Campbell, A. and Converse, P.E. (1972) *The Human Meaning of Social Change*, Russel Sage Foundation, New York.

Carley, M. (1981) *Social Measurement and Social Indicators: Issues of Policy and Theory*, George Allen & Unwin, London.

Champion, A.G. and Townsend, A. (1990) *Contemporary Britain: a Geographical Perspective*, Edward Arnold, London.

Cheshire, P.C. and Hey, D.G. (1988) *Urban Problems in Western Europe*, Unwin Hyman, London.

Chisholm, M. (1971) 'In search of a basis for locational theory: microeconomics or welfare economics?', *Progress in Geography*, 111–33.

Cini, M. (1990) 'Trentatre Variazioni su una Tema. Soggetti dentro e fuori la Scienza', Editori Riuniti, Rome.

CIPFA (1979) 'Community indicators', Chartered Institute of Public Finance and Accounting, London.

Clarke, G.P. (1986) 'Modelling retail centre size and location', Working Paper 482, School of Geography, University of Leeds.

Clarke, G.P. and Wilson, A.G. (1987) 'Performance indicators within a model-based approach to urban planning', *Sistemi Urbani, 2/3*, 137–65.

Clarke, M. and Holm, E. (1987) 'Microsimulation methods in spatial analysis and planning', *Geografiska Annaler, 69B*, 145–64.

Clarke, M. and Sinclair, T. (1991) 'Predictive geographical intelligence systems for retail planning and analysis: TOYOTA GB – a case study', in *Mapping Awareness 1992*, Conference proceedings, Blenheim, London.

Clarke, M. and Wilson, A.G. (1984) 'Modelling for health services planning', in Clarke, M. (ed.) *Planning and Analysis in Health Care Systems*, Pion, London, 22–56.

Clarke, M. and Wilson, A.G. (1985a) 'The dynamics of urban spatial structure', *Transactions, Institute of British Geographers, 10*, 427–51.

Clarke, M. and Wilson, A.G. (1985b) 'Developments in planning models for health care policy analysis in the U.K.', Working Paper 422, School of Geography, University of Leeds.

Clarke, M. and Wilson, A.G. (1986) 'A framework for dynamic comprehensive urban models: the integration of accounting and micro-simulation approaches', *Sistemi Urbani, 5*, 145–77.

Clarke, M. and Wilson, A.G. (1987) 'Towards an applicable human geography: some developments and observations', *Environment and Planning A, 19*, 1525–41.

Clarke, M., Keys, P. and Williams, H.C.W.L. (1980) 'Micro-simulation in socio-economic and public policy analysis', in Voogd H. (ed.) *Strategic Planning in a Dynamic Society*, Delft University Press, Delft.

Clarke, M., Keys, P. and Williams, H.C.W.L. (1981) 'Micro-simulation', in Wrigley, N. and Bennett, R.J. (eds) *Quantitative Geography*, Routledge & Kegan Paul, London, 248–56.

Coates, B.E., Johnston, R.J. and Knox, P.L. (1977) *Geography and Inequality*, Oxford

University Press, Oxford.
Coelho, J.D. and Wilson, A.G. (1976) 'The optimum location and size of shopping centres', *Regional Studies, 10*, 413–21.
Cohen, W.J. (1967) 'Education and learning', *Annals of the American Academy of Political and Social Science, 373*, 79--101.
Cooke, P. (ed.) (1989) *Localities: the Changing Face of Urban Britain*, Unwin Hyman, London.
Coombs, P.H. (1969) 'Time for a change in strategy', in Beeby, C.E. (ed.) *Qualitative Aspects of Educational Planning*, UNESCO, Paris, 39–70.
Cosgrove, D. (1989) 'Models, description and imagination in geography', in MacMillan, B. (ed.) *Remodelling Geography*, Basil Blackwell, Oxford, 230–44.
Craig, J. and Driver, A. (1972) 'The identification and comparison of all areas of adverse social conditions', *Applied Statistics, 2*, 25–35.
Cripps, E. and Foot, D. (1968) 'Evaluating alternative strategies', *Official Architecture and Planning, 31*, 928–38.
Cripps, E. and Foot, D. (1970) 'The urbanisation effects of a third London Airport', *Environment and Planning, 2*, 153–92.
Culyer, A.J. (1977) 'The quality of life and limits of cost–benefit analysis', in Wingo, L. and Evans, A. (eds) *Public Economics and the Quality of Life*, Johns Hopkins University Press, Baltimore, Md., 141–53.
Davidson, K.B. (1977) 'Accessibility in transport land-use modelling and assessment', *Environment and Planning A, 9*, 1401–16.
Davies, H. (1984) '1981 Census – a ward index of deprivation', Greater London Council, Statistical Series no. 35, London.
Davies, W.K.D. (1970) 'Approaches to urban geography: an overview', in Carter, H. and Davies, W.K.D. (eds) *Urban Essays: Studies in the Geography of Wales*, Longman, London, 1–21.
Denver Urban Observatory (1973) 'Urban social indicators: selected conditions and trends in Denver and its metropolitan area', Denver Urban Observatory, Denver.
Department of the Environment (1975) 'The use of indicators for area action – Housing Act 1974', Area Improvement Note 10, HMSO, London.
Detragiache, A. (1991) 'La citta e l'urbanistica', Quaderni Eidon, SGE, Naples.
DHSS (1972) 'Management arrangements for the re-organised National Health Service', HMSO, London.
DHSS (1982) 'Performance indicators in the N.H.S.: a progress report on the joint exercise between D.H.S.S. and Northern Region, DHSS, RA(82)34, London.
DHSS (1984) 'Performance indicators: national summary for 1981', Regional Liaison Division, London.
Dickinson, J.C., Gray, R.J. and Smith, D.M. (1972) 'The quality of life in Gainsville, Florida: an application of territorial social indicators', *Southeastern Geographer, 12*, 121–32.
Domencich, T. and McFadden, T. (1975) *Urban Travel Demand: a Behavioural Analysis*, North-Holland, Amsterdam.
Donna Bianco, P.A., Gaietta, G., Giudice, M., Rabino, G.A., Rosa, G. and Socco, C. (1981) 'A structural model for regional planning: the case study of Piedmont', in Lierop, Van W. and Nijkamp, P. (eds) *Locational Developments and Urban Planning*, Sijthoff of Noordhoff, Alphen an den Rijn, The Netherlands, 507–31.
Duncan, S.S. (1974) 'Cosmetic planning or social engineering? Improvement grants and improvement areas in Huddersfield', *Area, 6*, 259–71.
Ellickson, B. (1981) 'An alternative test of the hedonic theory of housing markets', *Journal of Urban Economics, 9*, 56–79.
Ferris, A. (1969) *Indicators of Trends in American Education*, Russell Sage Foundation, New York.
Fienberg, S.E. (1970) 'An iterative procedure for estimation in contingency tables',

Annals of Mathematical Statistics, *41*, 907–17.
Fitzsimmons, S.J. and Levy, W.G. (1976) 'Social economic accounts system (SEAS) toward comprehensive community-level assessment procedure', *Social Indicators Research, 2*, 389–452.
Flax, M.J. (1978) *Survey of Urban Indicator Data*, The Urban Institute, Washington, D.C.
Foot, D. (1981) *Operational Urban Models*, Methuen, London.
Fotheringham, S. and O'Kelly, M. (1988) *Spatial Interaction Models: Formulations and Applications*, Kluwer Academic Press, New York.
Fox, K.A. (1974) *Social Indicators and Social Theory: Elements of an Operational System*, Wiley, New York.
Gallino, T., Occelli, S. and Tadei, R. (eds) (1990) 'Costruzione di un sistema interattivo per l'analisi e la pianificazione dei sistemi localizzazione-transporti', *Rapporto Finale*, PFT 1-CNR, Contratto n.86.01192.93, Dipartimento di Scienze e Tecniche per i Processi di Insediamento, Turin University, Turin.
Gehrmann, T. (1978) 'Valid empirical measurement of quality of life?', *Social Indicators Research, 5*, 73–110.
Giggs, J.A. (1970) 'Socially disorganised areas in Barry: a multivariate approach', in Carter, H. and Davies, W.K. (eds) *Urban Essays: Studies in the Geography of Wales*, Longman, London, 101–43.
Glennester, H. and Hatch, S. (1974) 'Positive discrimination and inequality', Fabian Research Series 3/4, Fabian Society, London.
Goldacre, M. and Griffin, K. (1983) 'Performance indicators: a commentary on the literature', Unit of Clinical Epidemiology, University of Oxford, mimeo.
Gordon, I.R. (1986) 'The structural element in regional unemployment', Paper presented to the Annual Conference of the Regional Science Association, British Section, University of Bristol, September.
Grichting, W.L. (1984) 'The meaning of social policy and social structure', *International Journal of Sociology and Social Policy, 4 (4)*, 16–37.
Gross, B.H. (1966) 'The state of the nation: social systems accounting', in Bauer, R.A. (ed.) *Social Indicators*, MIT Press, Cambridge, Mass.
Guillebaud, C.W. (1956) 'Report on the committee of enquiry into the cost of the National Health Service', Cmnd 9663, HMSO, London.
Guy, C.M. (1977) 'A method for examining and evaluating the impact of major retail developments upon existing shops and their users', *Environment and Planning A, 9*, 491–504.
Guy, C.M. (1983) 'The assessment of access to local shopping opportunities', *Environment and Planning B, 10*, 219–38.
Gwilliam, K. (1972) 'Economic evaluation of urban transport projects: the state of the art', *Transportation Planning and Technology, 1*, 123–42.
Haag, G. (1990) 'Master equations', in Bertuglia, C.S., Leonardi, G. and Wilson, A.G. (eds) *Urban Dynamics: Designing an Integrated Model*, Routledge, London, 69–83.
Hall, P. (1982) *Urban and Regional Planning*, Penguin, Harmondsworth.
Hamnett, C. (1979) 'Area-based explanations: a critical appraisal', in Herbert, D.T. and Smith, D.M. (eds) *Social Problems and the City*, Oxford University Press, Oxford, 243–60.
Hansen, W.G. (1959) 'How accessibility shapes land use', *Journal of the American Institute of Planners, 25*, 73–6.
Harris, B. and Wilson, A.G. (1978) 'Equilibrium values and dynamics of attractiveness terms in production-constrained spatial interaction models', *Environment and Planning A, 14*, 813–27.
Harvey, D. (1973) *Social Justice and the City*, Edward Arnold, London.
Harvey, D. (1989) 'From models to Marx: notes on the project to "remodel"

contemporary geography', in MacMillan B. (ed.) *Remodelling Geography*, Basil Blackwell, Oxford, 211–16.
Harvey, S., Clarke, M. and Wilson, A.G. (1985) 'The use of performance indicators and interactive planning systems to examine inpatient psychiatric care in the Yorkshire Regional Health Authority', Working Paper 444, School of Geography, University of Leeds.
Hasluck, C. (1987) *Urban Unemployment: Local Labour Markets and Employment Initiatives*, Longman, Harlow.
Hatry, H.P. *et al.* (1977) 'How effective are your community services? Procedures for monitoring the effectiveness of municipal services', The Urban Institute, Washington, D.C.
Hay, A. (1985) 'Scientific method in geography', in Johnston, R.J. (ed.) *The Future of Geography*, Methuen, London, 129–42.
Hayden, F.G. (1977) 'Toward a social welfare construct for social indicators', *American Journal of Economics and Sociology, 32*, 129–46.
Henderson-Stewart, D. (1990) 'Performance management and reviewing local government', in Cave, M., Kogan, M. and Smith, R. (eds) *Output and Performance Management in Government: the State of the Art*, Jessica Kingsley, London, 106–23.
Henriot, P.J. (1970) 'Political questions about social indicators', *Western Political Quarterly, 23*, 235–55.
Hill, M. (1968) 'A goals-achievement matrix for evaluating alternative plans', *Journal of the American Institute of Planners, 34*, 19–29.
Hill, M. (1973) 'Planning for multiple objectives: an approach to the evaluation of transportation plans', Regional Science Research Institute, Monograph 5, Philadelphia, Penn.
Hirschfield, A. (1986) 'Urban deprivation: selected aspects with special reference to Leeds', Ph.D dissertation, School of Geography, University of Leeds.
Hirschfield, A. and Rees, P.H. (1984) 'Datapac: a program for the extraction of Census and vital statistics for Leeds wards', Working Paper 392, School of Geography, University of Leeds.
Holterman, S. (1975) 'Areas of deprivation in Great Britain: an analysis of 1971 Census data', *Social Trends, 6*, 33–47.
Hotelling, H. (1929) 'Stability in competition', *Economic Journal*, 39, 41–57.
Ingram, D.R. (1971) 'The concept of accessibility: a search for an operational form', *Regional Studies, 5*, 101–7.
Irwin, S. and Wilson, A.G. (1985) 'Performance appraisal in education', Working Paper 413, School of Geography, University of Leeds.
Johnstone, J.N. (1978) *Education Systems: Approaches and Methods in their Evaluation*, Evaluation in Education, International Progress, vol. 2, Monograph 3.
Johnstone, J.N. (1981) *Indicators of Education Systems*, Kogan Page/UNESCO, London/Paris.
Jowett, P. and Rothwell, M. (1988) *Performance Indicators in the Public Sector*, Macmillan, Basingstoke.
Juster, F.T. and Land, K.C. (eds) (1981) *Social Accounting Systems: Essays on the State of the Art*, Academic Press, New York.
Kemp, P. (1979) 'Planning and development statistics 1978/79', *District Councils Review*, 62–4.
Kennedy, L.W., Northcott, H.C. and Kinzel, C. (1978) 'Subjective evaluation of well-being: problems and prospects', *Social Indicators Research, 5*, 457–74.
Klein, R. (1982) 'Performance, evaluation and the N.H.S.: a case study in conceptual perplexity and organisational complexity', *Public Administration, 60 (4)*, 385–407.

Knox, P.L. (1975) *Social Well-being: a Spatial Perspective*, Oxford University Press, Oxford.
Knox, P.L. (1976) 'Social priorities or social indicators: a survey approach', Occasional Paper 4, Department of Geography, University of Dundee.
Knox, P.L. (1978a) 'Territorial social indicators and area profiles: some cautionary observations', *Town Planning Review, 49 (1)*, 75–83.
Knox, P.L. (1978b) 'Measures of accessibility and social indicators: a note', *Social Indicators Research, 7*, 367–77.
Knox, P.L. (1982) *Urban Social Geography: an Introduction*, Longman, London.
Knox, P.L. (1985) 'Disadvantaged households and areas of deprivation: microdata from the 1981 Census of Scotland', *Environment and Planning A, 17*, 413–25.
Knudsen, D. and Fotheringham, A.S. (1986) 'Matrix comparison, goodness-of-fit, and spatial interaction modeling', *International Regional Science Review, 10 (2)*, 127–47.
Koening, J.G. (1975) 'A theory of urban accessibility', Paper presented to the PTRC Summer Annual Meeting, University of Warwick, July.
Koening, J.G. (1980) 'Indicators of urban accessibility: theory and application', *Transportation, 9*, 145–72.
Kuz, T.J. (1978) 'Quality of life, an objective and subjective variable analysis', *Regional Studies, 12*, 409–17.
Land, K.C. (1971) 'On the definition of social indicators', *American Sociologist, 6*, 322–5.
Land, K.C. and McMillen, M.M. (1981) 'Demographic accounts and the study of change, with applications to the post World War Two United States', in Juster, F.T. and Land, K.C. (eds) *Social Accounting Systems: Essays on the State of the Art*, Academic Press, New York, 242–306.
Land, K.C. and Spilerman, S. (eds) (1975) *Social Indicator Models*, Russell Sage Foundation, New York.
Lasswell, H.D. (1958) *Politics: Who Gets What, When and How*, World Publishing Company, Cleveland, Ohio.
Lee, D.B. (1973) 'Requiem for large scale models', *Journal of the American Institute of Planners, 39*, 163–78.
Lee, R. (1976) 'Public finance and urban economy: some comments on spatial reformism', *Antipode, 8 (1)*, 43–50.
Leigh, C.M. (ed.) (1991) *A Science Park System for Leeds*, The Yorkshire and Humberside Regional Research Observatory, The University of Leeds, Leeds.
Leonardi, G. (1983) 'The use of random utility theory in building location-allocation models', in Thisse, J.F. and Zoller, H.G. (eds) *Locational Analysis of Public Facilities*, North-Holland, Amsterdam, 357–83.
Leonardi, G. (1984) 'The structure of random utility models in the light of the asymptotic theory of extremes', in Florian, M. (ed.) *Transportation Planning Models*, North-Holland, Amsterdam, 107–33.
Leonardi, G. (1985) 'Asymptotic approximations of the assignment model with stochastic heterogeneity in the matching utilities', *Environment and Planning A, 17*, 1303–14.
Leonardi, G. (1990a) 'Stochastic extremal processes', in Bertuglia, C.S., Leonardi, G. and Wilson, A.G. (eds) *Urban Dynamics: Designing an Integrated Model*, Routledge, London, 84–140.
Leonardi, G. (1990b) 'Labour market 3: a stochastic assignment approach', in Bertuglia, C.S., Leonardi, G. and Wilson, A.G. (eds) *Urban Dynamics: Designing an Integrated Approach*, Routledge, London, 329–45.
Lerman, S.R and Kern, C.R. (1983) 'Hedonic theory, bid rents and willingness to pay: some extensions of Ellickson's results', *Journal of Urban Economics, 13*, 358–63.
Lewis, G.M. (1968) 'Levels of living in the Northeastern United States 1960: a new

approach to regional geography', *Transactions of the Institute of British Geographers, 45*, 11–37.
Lewis, J. and Townsend, A. (eds) (1989) *The North–South Divide: Regional Change in Britain in the 1980s*, Paul Champion, London.
Lewis, S. and Jones, J. (1990) 'The use of output and performance measures in Government departments', in Cave, M., Kogan, M. and Smith, R. (eds) *Output and Performance Measurement in Government: the State of the Art*, Jessica Kingsley, London, 39–58.
Lichfield, N. (1966) 'Cost–benefit analysis in town planning: a case study: Swanley, *Urban Studies, 3*, 215–49.
Lineberry, R.L. (1977) *Equality and Urban Policy: the Distribution of Municipal Public Services*, Sage Publications, Beverley Hills, Calif.
Little, A. and Mabey, C. (1972) 'An index for designation of educational priority areas', in Shonfield, A. and Shaw, S. (eds) *Social Indicators and Social Policy*, Heinemann, London, 67–93.
Liu Ben-Chieh (1975) *Quality of Life Indicators in U.S. Metropolitan Areas 1970*, US Government Printing Office, Washington, D.C.
Liu Ben-Chieh (1978) 'Variations in social quality of life indicators in medium metropolitan areas', *American Journal of Economics and Sociology, 37*, 211–60.
Logan, R.F.L., Ashley, J.S.A., Klein, R.E. and Robson, D.M. (1972) 'Dynamics of medical care: the Liverpool study into the use of hospital resources', London School of Hygiene and Tropical Medicine, Memoir 14, London.
Lombardo, S.T. and Rabino, G. (1983) 'Non-linear dynamic models for spatial interaction: the results of some empirical experiments', *Papers of the Regional Science Association, 55*, 83–101.
Lombardo, S.T., Pumain, D., Rabino, G.A., Saint-Julien, T. and Sanders, L. (1987) 'Comparing urban dynamic models: the unexpected differences in two similar models', *Sistemi Urbani, 9*, 213–28.
Louis, A.M. (1975) 'The worst American city', *Harper's*, January, 67–71.
Lowry, I.S. (1964) *A Model of Metropolis*, RM-4035-RC, Rand Corporation, Santa Monica, Calif.
MacLaren, A. (1981) 'Area-based positive discrimination and the distribution of well-being', *Transactions of the Institute of British Geographers, New Series no. 6*, 53–67.
Martin, D. and Williams, H.C.W.L. (1992) 'Market area analysis and accessibility to primary health care centres', *Environment and Planning A, 24*, 1009–19.
McLoughlin, J.B., Nix, C.K. and Foot, D.H.S. (1966) 'Regional shopping centres: a planning report on North West England: Part 2: a retail shopping model', University of Manchester School of Town and Country Planning.
McNamara, P. (1973) 'A social report for Metropolitan Albuquerque', Albuquerque Urban Observatory, Albuquerque.
Mela, A. and Preto, G. (1990) 'Alla ricerca della strategia perduta', in Curti, F. and Diappi, L. (eds) *Gerarchie e Reti di Citta: Tendenze e Politiche*, Angeli, Milan, 127–54.
Mela, A., Preto, G. and Rabino, G. (1987) 'Principles of spatial organization: a unifying model for regional systems', Paper presented at the 5th European Colloquium on Quantitative and Theoretical Geography, Bardonecchia, mimeo.
Miller, Z.L. (1973) *The Urbanisation of Modern America: a Brief History*, Harcourt, Brace, Jovanovich, New York.
Mishan, E.J. (1972) *Cost–Benefit Analysis*, George Allen & Unwin, London.
Mondale, W. (1967a) 'New tools for social progress', *Progressive, 31*, 28–31.
Mondale, W. (1967b) 'Some thoughts on stumbling into the future', *American Psychologist, 22*, 972–3.
Monti, L. (1975) 'Social indicators for Austin, Texas: a cluster analysis of census

tracts', Bureau of Business Research, Graduate School of Business, University of Texas at Austin.

Morris, J.M., Dumbley, P.L. and Wigan, M.R. (1979) 'Accessibility indicators for transport planning', *Transportation Research, 13A*, 91–109.

Moser, G.A. and Scott, W. (1961) *British Towns; a Statistical Study of their Social and Economic Differences*, London.

Moulden, M. and Bradford, M.G. (1984) 'Influences on educational attainment: the importance of local residential environment', *Environment and Planning A, 16 (1)*, 49–66.

Mullen, P. (1989) 'Health and the internal market: implications of the White Paper', Discussion Paper 25, Health Service Management Centre, University of Birmingham.

Mullen, P. (1990) 'Planning and internal markets', in Spurgeon, P. (ed.) *The Changing Face of the NHS in the 1990s*, Longman, Harlow, 18–37.

Murphy, T.P. (1980) *Urban Indicators*, Urban Studies Information Guide Series, vol. 10, Gale Research, Detroit, Mich.

National Academy of Sciences Social Science Research Council (1969) *The Behavioural and Social Sciences: Outlook and Needs*, Prentice-Hall, Englewood Cliffs, N.J.

National Goals Research Staff (1970) *Toward Balanced Growth: Quantity with Quality*, US Government Printing Office, Washington, D.C.

Neuburger, H. (1971) 'User benefit in the evaluation of transport and land use plans', *Journal of Transport Economics and Policy, 5*, 52–75.

Nijkamp, P. (1984) 'Information systems: a general introduction', in Nijkamp, P. and Rietveld, P. (eds) *Information Systems for Integrated Regional Planning*, North-Holland, Amsterdam, 3–34.

Oberg, S. (1976) *Methods for Describing Physical Accessibility to Supply Points*, Lund Series in Geography, B43, Lund.

Occelli, S. (1989) 'Studio di fattibilita per la realizzazione di indicatori socio-ambientali finalizzati per il sistema informativo territoriale della Regione Piemonte', IRES, Turin.

O'Loughlin, J. (1983) 'Spatial inequalities in Western cities: a comparison of North America and German urban areas', *Social Indicators Research, 13 (2)*, 185–212.

Openshaw, S. (1986) 'Modelling relevance', *Environment and Planning, A, 18*, 143–47.

Openshaw, S. (1989) 'Computer modelling in geography', in Macmillan B. (ed.) *Remodelling Geography*, Basil Blackwell, Oxford, 70–88.

Pacione, M. (1990) 'What about people? a critical analysis of urban policy in the U.K.', *Geography, 75 (3)*, 193–202.

Pollitt, C. (1985) 'Measuring performance: a new system for the National Health Service', *Policy and Politics, 13 (1)*, 1–15.

Pollitt, C. (1990) 'Performance indicators, root and branch', in Cave, M., Kogan, M. and Smith, R. (eds) *Output and Performance Measurement in Government: the State of the Art*, Jessica Kingsley, London, 39–58.

Pred, A. (1977) *City-Systems in Advanced Economies*, Hutchinson, London.

Rabino, G.A. (1991) 'L'evoluzione dell'ingegneria del territorio e dell'ambiente', Proceedings of the 36th National Congress of Italian Chartered Engineers, Como.

Rabino, G.A. (forthcoming) 'Cambiamento delle citta ed innovazione tecnologica. Fondamenti della teoria evoluzionista dei sistemi urbani', in Lombardo, S.T. (ed.) *Nuove Tecnologie dell'Informazione e Sistemi Urbani*, Nuova Italia Scientifica, Rome.

Rawls, J. (1973) 'Distributive justice', in Phelps, S. (ed.) *Economic Justice*, Penguin, Harmondsworth, 319–62.

Rhind, D. (1989) 'Computing, academic geography and the world outside', in

MacMillan, B. (ed.) *Remodelling Geography*, Basil Blackwell, Oxford.
Roberts, M. (1990) 'Performance and outcome measures in the Health Service', in Cave, M., Kogan, M. and Smith, R. (eds) *Output and Performance Measurement in Government: the State of the Art*, Jessica Kingsley, London, 86–105.
Rogerson, R.J., Findlay, A.M., Morris, A.S. and Coombes, M.G. (1989) 'Indicators of quality of life: some methodological issues', *Environment and Planning A, 21*, 1655–66.
Rowley, C.K. and Peacock, A.T. (1975) *Welfare Economics – a Liberal Restatement*, Wiley, London.
Samuelson, P.A. (1973) *Economics*, McGraw-Hill, New York.
Sayer, R.A. (1976) 'A critique of urban modelling', *Progress in Planning, 6*, 187–254.
Schneider, J.B. and Symons, J.G. (1971) 'Regional health facility system planning: an access opportunity approach', Regional Science Research Institute, Discussion Paper 48, Philadelphia, Penn.
Schneider, M. (1974) 'The quality of life in large American cities: objective and subjective indicators', *Social Indicators Research, 1*, 495–509.
SELNEC Transportation Study (1972) 'A broad plan for 1984', Town Hall, Manchester.
Sheldon, E. and Moore, W.E. (eds) (1968) *Indicators of Social Change: Concepts and Measurements*, Russell Sage Foundation, New York.
Sheppard, E. (1987) 'A Marxian model of the geography of production and transportation in urban and regional systems', in Bertuglia, C.S., Leonardi, G., Occelli, S., Rabino, G., Tadei, R. and Wilson, A.G. (eds) *Urban Systems: Contemporary Approaches to Modelling*, Croom Helm, London, 189–250.
Shevky, E. and Bell, W. (1955) *Social Area Analysis*, Stanford University Press, Stanford, Calif.
Shevky, E. and Williams, M. (1949) *The Social Areas of Los Angeles*, University of California Press, Los Angeles, Calif.
Shonfield, A. and Shaw, S. (eds) (1972) *Social Indicators and Social Policy*, Heinemann, London.
Smith, D.M. (1973) *The Geography of Social Well-being in the United States: an Introduction to Territorial Social Indicators*, McGraw-Hill, New York.
Smith, D.M. (1974) 'Who gets what where and how: a welfare focus for human geography', *Geography, 59*, 289–97.
Smith, D.M. (1977) *Human Geography: A Welfare Approach*, Edward Arnold, London.
Smith, D.M. (1979) *Where the Grass is Greener: Geographical Perspectives on Inequality*, Croom Helm, London.
Smith, D.M. and Gray, R.J. (1972) 'Social indicators for Tampa, Florida', Urban Studies Bureau, University of Florida, mimeo.
Stegman, M.A. (1979) 'Neighbourhood classification and the role of the planner in seriously distressed communities', *Journal of the American Planning Association, 45*, 494–505.
Stipak, B. (1979) 'Citizen satisfaction with urban services: potential misuse as a performance indicator', *Public Administration Review, 39*, 46–52.
Stone, R. (1971) *Demographic Accounting and Model Building*, OECD, Paris.
Stone, R. (1975) 'Transitions and admission models in social indicator analysis', in Land, K.C. and Spilerman, S. (eds) *Social Indicator Models*, Russell Sage Foundation, New York, 253–300.
Sullivan, J.L. (1971) 'Multiple indicators and complex causal models', in Blalock, H.M. (ed.) *Causal Models in the Social Sciences*, Aldine, Chicago, Ill., 327–34.
Sullivan, J.L. (1974) 'Multiple indicators: some criteria of selection', in Blalock, H.M. (ed.) *Measurement in the Social Sciences*, Aldine, Chicago, Ill., 243–69.

Tanner, J.C. (1980) 'Distribution models for the journey to work', Transport and Road Research Laboratory Report 951, TRRL, Crowthorne, Berkshire.
Thompson, E.J. (1978) 'Social trends: the development of an annual report for the United Kingdom', *International Social Science Journal, 30*, 653–9.
Todd, R.H. (1977) 'A city index: measurement of a city's attractiveness', *Review of Applied Urban Research, 5*, 1–16.
Train, K.E. and McFadden, D. (1978) 'The goods/leisure trade off and disaggregate worktrip mode choice models', *Transportation Research 12 (5)*, 349–53.
Treasury (1961) 'The financial and economic obligations of the nationalised industries', Cmnd 1337, HMSO, London.
Treasury (1967) 'Nationalised industries: a review of economic and financial objectives', Cmnd 3437, HMSO, London.
Tressider, J.O., Meyers, D.A., Burrell, J.E. and Powell, T.J. (1968) 'The London Transportation Study: methods and techniques', *Proceedings of the Institution of Civil Engineers, 39*, 433–64.
Urban Observatory of San Diego (1973) *The Quality of Life in San Diego: Selected Indicators of Urban Conditions and Trends 1973*, Urban Observatory of San Diego, San Diego, Calif.
US Department of Health, Education and Welfare (1969) *Toward a Social Report*, US Government Printing Office, Washington, D.C.
US Office of Management and Budget (1973) *Social Indicators 1973*, US Government Printing Office, Washington, D.C.
US Office of Management and Budget (1976) *Social Indicators 1976*, US Government Printing Office, Washington, D.C.
Van den Berg, L., Drewett, R., Klassen, L.H., Rossi, A. and Vijverberg, C.H.T. (1982) *Urban Europe*, vol 1: *A Study of Growth and Decline*, Pergamon, Oxford.
Van Dijk, J. and Folmer, H. (1985) 'Entry of the unemployed into employment: theory, methodology and Dutch experience', *Regional Studies, 19 (3)*, 243–56.
Voogd, H. (1983) *Multicriteria Evaluation for Urban and Regional Planning*, Pion, London.
Vos, J.B., Feenstra, J.F., de Boer, J., Braat, L.C. and van Baalen, J. (1985) *Indicators for the State of the Environment*, Institute of Environmental Studies, Free University of Amsterdam, The Netherlands.
Waddington, C.H. (1977) *Strumenti per Pensare: Un Approccio Globale ai Sistemi Complessi*, Mondadori, Milan.
Walker, B. (1980) 'Urban planning and social welfare', *Environment and Planning A, 12*, 217–25.
Warren, R.D., Fear, F.A. and Klonglan, G.E. (1980) 'Social indicator model building: a multiple-indicator design', *Social Indicators Research, 7*, 269–97.
Webber, M.J. (1987) 'Quantitative measurement of some Marxist categories', *Environment and Planning A, 19 (10)*, 1303–22.
Webber, R.J. (1977) *The National Classification of Residential Neighbourhoods: an Introduction to the Classification of Wards and Parishes*. Town Planning Research Applications Group, London.
Webber, R.J. (1978) 'Making the most of the Census for strategic analysis', *Town Planning Review, 49*, 274–84.
Webber, R.J. (1979) *Census Enumeration Districts: a Socio-economic Classification*, OPCS, London.
Weibull, J.W. (1976) 'An axiomatic approach to the measurement of accessibility', *Regional Science and Urban Economics, 6*, 357–79.
Weidlich, W. and Haag, G. (1983) *Concepts and Models of a Quantitative Sociology. The Dynamics of Interacting Population*, Springer Series in Synergetics 14, Springer, Berlin.
Williams, H.C.W.L. (1976) 'Travel demand models, duality relations and user benefit

analysis', *Journal of Regional Science, 16*, 147–66.

Williams, H.C.W.L. (1977) 'On the formation of travel demand models and economic evaluation measures of user benefit', *Environment and Planning A, 9*, 285–344.

Williams, H.C.W.L. and Senior, M.L. (1978) 'Accessibility, spatial interaction and spatial benefit : analysis of land use–transportation plans', in Karlquist, A., Lundquist, L., Snickars, F. and Weibull, J.W. (eds) *Spatial Interaction Theory and Planning Models*, North-Holland, Amsterdam , 253–88.

Williams, P. and Fotheringham, A.S. (1984) *The Calibration of Spatial Interaction Models by Maximum Likelihood Estimation with Program SIMODEL*, Geographic Monograph Series Volume 7, Department of Geography, Indiana University.

Willms, J.D. (1992) *Monitoring School Performance: a Guide for Educators*, Falmer Press, London.

Wilson, A.G. (1967) 'A statistical theory of spatial distribution models', *Transportation Research, 1*, 253–69.

Wilson, A.G. (1970) *Entropy in Urban and Regional Modelling*, Pion, London.

Wilson, A.G. (1971) 'A family of spatial interaction models and associated developments', *Environment and Planning, 3*, 1–32.

Wilson, A.G. (1974) *Urban and Regional Models in Geography and Planning*, Wiley, London.

Wilson, A.G. (1976) 'Retailers' profits and consumers' welfare in a spatial interaction shopping model', in Masser, I. (ed.) *Theory and Practice in Regional Science*, Pion, London, 42–59.

Wilson, A.G. (1981a) *Catastrophe Theory and Bifurcation: Applications to Urban and Regional Models*, Croom Helm, London.

Wilson, A.G. (1981b) *Geography and the Environment: Systems Analytical Methods*, Wiley, Chichester.

Wilson, A.G. (1984) 'Making urban models more realistic: some strategies for future research', *Environment and Planning A, 16 (11)*, 1419–32.

Wilson, A.G. (1988) 'Store and shopping centre location size: a review of British research and practice', in Wrigley, N. (ed.) *Store Choice, Store Location and Market Analysis*, Routledge, London, 160–86.

Wilson, A.G. (1989) 'Classics, modelling and critical theory: human geography as structured pluralism', in MacMillan, B. (ed.) *Remodelling Geography*, Basil Blackwell, Oxford, 61–9.

Wilson, A.G. and Birkin, M. (1987) 'Dynamic models of agricultural location in a spatial interaction framework', *Geographical Analysis, 19 (1)*, 31–56.

Wilson, A.G. and Crouchley, R. (1984) 'The optimum sizes and locations of schools', Working Paper 369, School of Geography, University of Leeds.

Wilson, A.G., Coelho, J.D., Macgill, S.M. and Williams, H.C.W.L. (1981) *Optimization in Locational and Transport Analysis*, Wiley, London.

Women and Geography Study Group (1984) *An Introduction to Feminist Geography*, Hutchinson, London.

Worrall, L. (1986) 'Information systems for urban labour market planning and analysis', Paper presented to the Annual Conference of the Regional Science Association, British Section, September.

Yates, J. (1982) *Hospital Beds. A Problem for Diagnosis and Management?*, Heinemann, London.

Index